Career Management for Chemists

T0190502

Springer
Berlin
Heidelberg
New York
Hong Kong
London
Milan
Paris
Tokyo

John Fetzer

Career Management for Chemists

A Guide to Success in a Chemistry Career

 Springer

Dr. John Fetzer
Fetzpahs Consulting
94564 Pinole
CA, USA

fetzpahs@hotmail.com

ISBN 978-3-642-05886-8

Cataloging-in-Publication Data applied for

Bibliographic information published by Die Deutsche Bibliothek
Die Deutsche Bibliothek lists this publication in the Deutsche Nationalbibliografie;
detailed bibliographic data is available in the Internet at <http://dnb.ddb.de>.

Springer-Verlag is a part of Springer Science+Business Media
© Springer-Verlag Berlin Heidelberg New York 2010
Printed in Germany
springeronline.com

Coverdesign: Erich Kirchner, Heidelberg
52/3020 uw   Printed on acid-free paper – 5 4 3 2 1 0 –

Preface

The writing of this book was a broad range of experiences. I started out thinking it would be an expansion of my monthly essays in Analytical and Bioanalytical Chemistry, just being more detailed and writing more on topics. Instead, I soon realized that the topics needed were many and more varied. Even within an area that I had already covered in one of the published essays, I found through thinking, reading, and discussing with other scientists that there were more facets. Topics such as listening to communicate, ethics, the dynamics of teams, and the cultural aspects of science became topics of their own.

The topics are wide in scope and hopefully helpful in many ways to the wide variety of scientists who read it for ideas.

Through working with the illustrator, Kevin Coffey of Cartoonlandanimation.com, some of the ideas became visual and not just concepts. This was an interesting exercise in itself as I believe that there is a lot of aesthetic beauty in science anyway.

I am interested in any new topics or perspectives, if not for some future second edition, then for new essay topics.

Good luck to each reader in finding steps towards success.

John Fetzer

Contents

Acknowledgements

Many people helped in forming this book. Besides the many people I worked with over the years who helped create my experiences and ideas, there are those who I spoke with on career development for research chemists for this book. Christina Dyllick, editor of Analytical and Bioanalytical Chemistry, let me create the essays that generated the book. In that phase, two of the journal's editors, Sylvia Daunert and Philippe Garrigues, came up with several good ideas for essays. Douglas Lane, Jocelyn Hellou, Kathleen Kilway, Linda Osbourne, Richard Mathies, Wilton Biggs, and Stephen Wise gave especially useful and extensive comments and ideas.

I would also like to thank the illustrator, Kevin Coffey. It was very nice exchanging ideas and seeing some of the writing turned into visuals. I hope you, the readers, enjoy them, too.

1 Introduction – The Career as a Long Trip

1   Introduction – The Career as a Long Trip

While writing this book, someone suggested to me that I write the sections of it as if an automobile trip was an analogy to a scientific research career. There are many similarities. You have to prepare yourself for the trip, gather up what you'll need for the trip, pack them in, find maps, and the many other preparations. These could be analogs to gathering the right attitudes and skills to do research. Once on the journey, you ask for directions, you stop to eat and refuel your vehicle, and you may take interesting side trips or even change your destination as you travel. These are analogous to being mentored, and adding to your knowledge to become both expert and diverse.

The more I thought of this analogy, however, the better I thought of it as a slower-paced trip. The ideas transformed into a journey by a different conveyance, maybe a combination of a walking tour with some stretches traveled by train. The pace is slower and more flexible, as is research. You meet more people while walking, sometimes even traveling together. This is analogous to collaborating and sharing new discoveries. A research career is a long journey with innumerable choices and possibilities, just as wandering on foot through interesting places can be. You do not need to follow a guide or guidebook, but you can. You wander as your interests take you and find new and wonderful things. Each person creates an own tour with an own itinerary.

The origins of this book are found in my writing a series of monthly essays for the journal Analytical and Bioanalytical Chemistry. The series arose after I wrote a guest editorial for the journal. I had written on my experiences as an industrial chemist and how graduate school had not prepared me for the differences in the environment and situations compared to academia. This turned into an exchange with the editor, Christina Dyllick, about extending that one essay into more.

4

This essay series was defined as each essay dealing with some area of the non-technical aspects of analytical chemistry. Their broader application to the other disciplines of chemistry and even to the other experimental sciences seemed to be a natural extension. Naturally many of the topics dealt with how to do good science and be successful. If some of the points in this book seem to be sermonizing, this comes from both the fact that the origins of some parts of the chapters were those essays with their editorial tone and also because I strongly believe in these ideas. This is especially true for the emphasis on the human sides of science, the need for scientists to understand the psychology and sociology of our profession. If any reader has seen those essays, then some of these chapters will seem familiar. I used the first couple of dozen essays as the basis for some of the chapters. They were extended, elaborated, and added to with more details and other ideas.

I must admit that some of my motivation for both the essay series and for this book comes from my passion for science. I knew at an early age, maybe around when I was ten years old, that I wanted to be a research scientist. By my middle teenage years, I knew this would be in chemistry. The rest of the next decade was making this happen. Helping others do well in their careers is part of this continuing love for science, discovery, and learning. I have always tried to make my ideas, whether technical or otherwise, a resource for others. I hope some of my enthusiasm shows through these chapters and gives insights in your career paths.

Some readers may feel that there is an over-emphasis on these aspects in this book, what some call the "touchy feely" sides. This may be so, but it is intentional that these get strongly emphasized. The part of research that deals with science and technology get an inherent emphasis in graduate schools and in the later careers. There often is the perception that these are all that matter on the road to success as a researcher. Although this may be true in a few cases in which scientific brilliance is so great that it alone is sufficient, in the much larger proportion of times good science skills augmented by people skills will be the key. As an example of the importance of the skills highlighted in this book, the April 21, 2003 Chemical and Engineering News carried an article on the characteristics biotechnology companies look rating and networking as one example. If at times there seems to be redundancy it is more due to these similarities and should be taken as a re-emphasis that the ideas are important in a very wide range of career tasks.

There are other books and papers that deal with some of these topics. These areas are being widely discussed and described in many books, magazine articles, Internet websites, and other sources; a few are listed in the bibliography at the end of the book. I have tried to include both the few sources that are written for chemists and those general books and articles that are also useful in many aspects of life besides chemistry careers. These include books on communication, self-assessment, positive and forward thinking, organizing and planning, and other areas. Some of those books are very well known, while others are more obscure. I tried to list sources that are readily available through libraries and booksellers. Use them as well, if that is your need. Find your own set of working practices.

For myself, two have been particularly useful, "Career Transitions for Chemists" by Dorothy Rodman, Donald Bly, Fred Owens, and Ann-Claire Anderson and "Career Management for Scientists and Engineers" by John K. Burchardt. The first is aimed specifically at people wanting to change career areas, while the second is more wide-ranging.

The first book deals with the issues in finding and landing a job. It touches on some areas that are not covered in this book such as employment opportunities within the various disciplines of chemistry. Some of the content describing the various chemical fields, however, is written for a limited audience, those who do not have expertise in hand and who can readily change career areas. This means more for someone choosing these areas either as a non-advanced degree chemist or as someone willing to totally change fields where their current expertise is not marketable.

The second book describes skill building in order to make yourself valuable and marketable. Additionally, it touches on more of the mechanics of job seeking such as interviewing and negotiating with a prospective employer.

Although I refer and relied on numerous sources, much of the observations and comments are my own. I naturally observe people and try to understand their thinking. This has

for in new hires. They quote one large company's
medicinal research "They (referring to chemists v
need excelle.t communications skills and the abili
contributions across disciplines." These are non-t

In some of those sections that came dir
essays, I focus more on my own research experier
issues in my core area of knowledge, analytical ch
points, however, are applicable to all research are
themes are that there are no boundaries between
fields and disciplines; moving from one research
is a key to remaining dynamic; thinking must ent
and skepticism; and that there are more similariti
differences in the various research venues.

I have striven to deal with several caree
treatments and discussions I have offered here we
the humanistic side of science in mind. Each scier
with thoughts, emotions, and attitudes that reflect
personality. Although there are a few other books
some of the topics here, I hope that this one adds
aspect which I believe can be the key to a successf
scientist understands the attitudes and thinking t
the chances of success, then not only will better sc
and more opportunities will be taken, but doing tl
more enjoyable and fulfilling.

Many of the ideas given here are my ow
personal experiences. I, however, have also includ
other chemists have given me in numerous interv
conversations, and exchanges of correspondence.
person deals with the tasks and issues must vary s
different person. Personal preferences are most in
defining the ways scientists do their work. Use thc
work for you. Meld and vary them if the result is e
own needs. This book is a guide, but not a precise
everyone needs different ways of doing things tha
personalities and ways of thinking.

A lot of the concepts are interconnected.
many of the elements of success involve the same
attitudes, emotions, ways of thinking, and ways of
Being a good communicator, being personable, be
being reliable and other aspects are both involved

helped me in relating to them. Understanding a person's motivations and how the thinking goes on makes it easier to interact. For my many friends and colleagues, here are a few words. If the person described is in a positive light and sounds familiar, it obviously must be you. If it is in a negative light, it obviously is someone else. Overall, I thank all of you for letting me learn more about you and the workings of research.

To some the ideas of understanding other people's personalities and attitudes and understanding how people interact with each other might seem to be unnecessary or even unpleasant. It has been my experience that science is only another facet of life and people behave similarly in it as they do in other things. If you ignore that, you miss opportunities. Some might think that using those understandings is manipulative or even Machiavellian. That would only be true if the motivations were solely for an individual's gain without any thought for the others involved. I think of it as using psychological and sociological principles to increase everyone's opportunities. If you open up more good possibilities for those others, then you are not manipulating them in the pejorative sense that word can have.

Overview – General Considerations

The focus of this book is scientific careers, particularly ones in research and focused on my own field of chemistry. Many of the topics will have specific insights on career-related matters for scientists. There, however, also is an underlying theme in many that are ideas that are relevant to any career area. In this section I will describe some of these. This is an initial attempt to emphasize many important ideas that are true in general whether you are a research chemist in academi!, industry, or a government laboratory, or someone working in management, chemical information, science writing, patent law, regulatory affairs, or the other diverse areas open to chemists.

The first and one of the most important points is that each person, no matter what their current position might be, must remember that they are really working for themselves. What is meant by this? Although many people may be interested in your career, only you control and shape it. You gain or lose by the way you manage it. Being aware of this is absolutely necessary in the current climate of cutbacks by industry, government laboratories, and even academia. Every person cannot rely on their employer for security. There are no guarantees. A lifetime employment is no longer the norm in industry. It is much less so in government and even an issue in academia. In the past, an individual could work in the same place for an entire career. Only those seeking higher salaries, more challenges, more prestige, and other personal gains switched positions often. This was a personal choice. In today's world, all that has changed. Maintaining one's marketability is a constant requirement for the long-term aim of a successful career.

Today these changes are often made without the individual's involvement. Mergers of companies can result in closures of laboratories. Budget cutbacks or redistributions can eliminate positions. These often come with little or no forewarning. One must always be prepared. No matter what a person's situation, she or he must constantly be aware of this. What are the most useful things to do in preparation for an unexpected career change? Networking and keeping an up-to-date resume or curriculum vitae are the two most useful answers. These will each be discussed in detail later in separate sections.

Not only are these two measures good in the case of unexpected career changes, but they also increase the chances for greater career opportunities in any situation. A better position that might go unnoticed can be found by having a good network of colleagues who know your interests and career wishes. Job opportunities often arise without wide knowledge or advertising. Word of it is spread only through chains of a few contacts connecting in sequence. If dozens of your colleagues are aware of your aims, then the chances are greater that one of them will hear of a good opportunity for you.

In keeping an up-to-date resume or CV, one prepares for those opportunities that have short timeframes or in which immediate response may make the difference in being considered. If not, then the time to put one together may be just enough for other people to submit theirs and be chosen. First impressions are important, both in the one's you make and in being the first to make an impression.

If one is open to a variety of positions and has the expertise to do so, it is advantageous to have more than one resume. Each version is written so that it targets positions in a particular area. For example, someone looking for an industrial position may have versions written for chemical, biotechnology, and environmental chemistry positions. Each highlights accomplishments in those areas and can leave out others that would be irrelevant for the others. The same is true if one can do research, supervise a lab, or manage a group. In academia, a resume or CV might emphasize teaching or research experiences depending on the position being aimed for.

9

This touches on another aspect of today's job market. Flexibility and versatility lead to both being able to respond quickly to new challenges and to moving into new areas. Th)s refers to both an overall career planning orientation and also to any opportunities within a current position. Making oneself more valuable through expanded expertise and responsibilities covers both aspects. This may not make a difference if the elimination of positions is severe, but it may make a difference if choices have to be made between individuals by managers. In addition, being diverse in interests and capabilities brings the personal rewards of maintaining dynamic research. Science has few real boundaries and the areas which are hot change with time.

Keeping up with developments in one's current areas of technical responsibility, as well as learning new ones that are of potential use, help in this flexibility and versatility. Always being in a learning mode must be part of a person's career plan in every situation.

The Personality in Science

In the different chapters I will deal with specific topics. Since one of the aims of this book is to treat science careers as an extension of the people who are the scientists, there will be much emphasis on personality and attitudes. I think these points will make this book a uniquely valuable guide since other career development books deal only with the mechanics of tasks to do and how they can be done effectively. To me, this is only a part of what goes into the building of a successful career.

Another major theme in this book is that scientists are people. Understanding your own thinking and emotions and those of other scientists that you may interact with will be helpful in your career. I describe a personality assessment, the Myer-Briggs Personality Type Indicator (MBPTI), in the chapter on personal assessments and matching career, job, and tasks to it. The MBPTI, however, is a useful tool and concept in understanding networking, diversity, collaboration, teamwork, success as a supervisor or manager, and other topics. Describing it in its most logical place does not mean its uses are bound to that area alone. The second most important use of the MBPTI is in pointing out the differences in people, learning to accept those differences, and then expanding your way of working with others to allow that diversity to be an asset. There is a very large, inherent bias with each of us. We think "I am a normal person, so any other normal person must think as I do." This is totally wrong! There is a huge diversity in thinking.

For example, some people work best by themselves. Collaboration for them is dividing up a project into individual tasks that are spread out among other individuals. Each task which that person does would be solo, while the others involved in the project each do their own tasks. This is in contrast to those who can work together on the same tasks.

2   Technical Areas

2 Technical Areas

2.1 Accepting Failure to Create Innovation in Experimentation

One of the most important lessons that a young scientist must learn is that good and innovative research is a delicate balance of many failures and few successes. This may not be the perception that some people have of experimental science because they only read research papers and hear presentations that highlight the successes. On occasion there are brief, passing references in these to things tried which failed, but the overall impression is that successes vastly outnumber failures. This perception of research success also adds weight to a false perception that many young scientists have.

If research is truly innovative, this cannot be so. If research is well enough understood to be always accurately predictable, then it can only be either "filling in the blanks" or slightly extending the boundary of well-established knowledge.

The first is replicating the research of others and looking more closely at the variations in experimental conditions. This includes not only variation of temperatures, reagent concentrations, and other physical variables, but also applying a technique to similar sample types.

The second is slightly more exploratory, but is not true experimentation if the results can routinely be predicted. Both types of work are necessary, overall, in order to fully understand the science but neither is very creative, very innovative, or highly challenging. A good research career should be a blend of both these types of work and ones involving more risk-taking. Innovation should be synonymous with risk-taking.

One of the most important lessons that a beginning graduate school student must learn is that experiments fail and very often do. The same is true for the analogous beginning scientist with an undergraduate degree in a laboratory position in industry. Their backgrounds are based on simple, well-defined, and well-understood "experiments". The laboratory exercises in an undergraduate analytical or organic course are graded on the basis of things like gravimetric yields, results of compositional analyses, intermediary and overall yields of a sequential organic synthesis, and other similar tasks.

These are all designed to build lab skills and teach meticulousness. The aims are known and success in results is defined by the closeness to ideal values. This mentality gets continually reinforced through the laboratory course work. The students do not understand that the purpose is mainly to teach good laboratory practices of care, thoroughness, and documentation. The goal, from a career standpoint, is not hitting the target values, but learning how to hit them. In doing so, the student builds lab skills that will be useful in research and any other laboratory work.

Real experimental science is fundamentally different. No matter how well understood the theories leading up to the experiments are or how well-designed those experiments are or how carefully the experiments are done, the end result often is nothing like what was expected. The results can be thought of as failures or as a learning that the plan was based on an unknown flaw. Experimental science delves into the unknown, so the work beforehand is a best guess at what might be. Sometimes these best guesses end up being totally wrong and the series of experiments yield nothing other than the fact that there is something unexplained.

I have often seen a young scientist frustrated or in an emotional turmoil because an experiment they had worked on had not worked as expected. A more experienced scientist just accepts these setbacks as inherent to the work. The shift in attitudes often comes with time, but the psychological and emotional tolls are often much more than they should be. This is because the young, learning scientist does not know that there is that difference between learning exercises and experiments.

This mindset is particularly going to be at its strongest among the brightest and most capable of young scientists. In undergraduate school work, they were the ones more likely of getting yields of 63 % when 64 % is the best expected for an organic synthesis or getting 5.62 grams of bright white silver chloride in a gravimetric exercise that should yield 5.65 grams. No off-color products or low yields were in their backgrounds. So the idea of experimental correctness sets in very strongly. Failure is tantamount to doing something wrong, not to doing it right. It is very difficult for these exceptional students to accept and understand that good science inherently is full of failed experiments. These people are also the ones with the most potential to become good researchers once they surmount this attitude. Getting them to accept failed experiments is good.

Getting young researchers to accept failure and assume the risks of stepping out into the areas of experimental unknowns is difficult. It also is very rewarding because the resulting researchers often are creative innovators. A more seasoned colleague, mentor, or professor (depending on the venue) must point out the fundamental difference between their experience so far (experiments that always end in the expected manner) and the reality of science (where most experiments are trials and errors; nuggets of new learning buried among much more barren rock).

On the contrary, if this learning never occurs, the laboratory scientist only does "research" on the creeping edges of knowledge. This takes many forms that can deceive the scientist into thinking that innovative and creative work is being done. In reality, this very predictable work with expected results is not research, but is only good laboratory work. Tweaking the variables for a fuller understanding is needed, but should not be the future of a career's work for a talented young scientist.

Research is exploratory. It delves into the unknown. By this it also must be a gamble with risk-taking. Resources and time are bet on the chances of a discovery. As with wagering on a race horse, betting on the favorites brings less return. Likely results are less major in the new information won by risking little in the experiments. Betting on long shots brings very large payments. This is just as experiments that leap far into the unknown by being based on many suppositions can be rewarded by learning many things unexpected. These are the major breakthroughs. Those are not the bulk of a researcher's work, though, because the efforts, time, and resources seldom can be sustained for the one-in-a-hundred. Most good researchers go for experimenting in between with occasional work in both the safer and riskier areas.

Without understanding the risk-taking that has been done, other researchers see the progress of those who were once peers and wonder why. They do not see the failures that accompany the innovative successes. They do not see the much more difficult efforts of doing several experiments as part of attaining the goal of a few successes. The difference between an innovator and the average scientist is often the acceptance of failures and of the risk-taking.

Over the course of time, a scientist's confidence builds so that these ideas and creative speculations become routinely thought of, used, and even exchanged among colleagues and peers. Many experienced researchers will tell you that one of the more interesting and entertaining aspects of conferences are the in the hallway or during coffee-break discussions on new ideas and possible experiments. These speculative discussions only happen when risk-taking is a normal part of one's thinking. Personally, I have become involved in several collaborations and created the frameworks for over a dozen papers in the course of such discussions.

The results not turning out as expected can be a challenge to the researcher. Why did it not turn out? If the researcher uses experimental failures for learning and challenges, then failed experiments become a part of an overall framework that sometimes leads to success (I use "sometimes" because we as scientists never can know how difficult the quest for knowledge can be. Our best efforts may be limited by the current theories and technologies).

If risk-taking and accepting failure are part of research, they also must be part of a loop in which curious and skeptical thinking are part. If an experiment does not result as you planned, you should obviously ask "Why?" This means investigations into the reasons that your best laid plans did not produce fruitful results. Reviewing the initial experiments for errors, reproducibility, and unknown variables are part of the skeptical thinking. Failed experiments are not just a reflection of risk-taking; they are also doors opening on new discoveries.

Richard Mathies, a bioanalytical chemist of the University of California at Berkeley, states emphatically that in his research unexpected results – the "failures" – have been much more important than those results that turned out as current theories predicted. These unexpected results created challenges and forced new innovative thinking because the accepted theories fail. They do not predict nor explain failed experiments that were planned using their premises... This often leads into experimental areas that would not have been looked at under the former theoretical models.

Dr. Mathies also points out that one paradigm in thinking must shift in our current system. Reporting of negative results is not common in the scientific literature. He contends that these help delineate both the limits of current theory and the path that new theory must move into. If these go unreported or only slightly reported in a spotty fashion, then researchers cannot readily come up with better theories and models to describe natural phenomena. The failures also define what new theories must explain.

There should be a tendency in science presentations, both written and spoken ones, to tell the story of the research. If this includes experiments that did not work, but which led to later success, then the story is complete and follows a logical path. This is especially needed when the failed paths were what seemed like the "obvious" route under the previous assumptions and theories. Even after a different successful course is discovered, other scientists may look at the work and think "Why did they not do this? It's obvious" and then proceed to redo the failures. Thus, by not reporting on the "obvious" course that failed, one scientist sets up others to do a wasted redundancy.

I am thinking of a failed experiment in a collaboration I was involved in. I had prepared a moderate amount of a specific compound, a few grams. This was probably the world's supply of this compound! An organic chemist asked me if there was any more available. He had a very good idea, or so it seemed from known mechanisms and behaviors, to make a certain target compound. This attempt failed. But he, his graduate students, and I are the only ones to know this "obvious" route fails. If there is a successful, subsequent synthesis of a compound, it should be common practice to report those attempts that did not work. Unfortunately, this is only done haphazardly in the literature.

On a brighter note, sometimes the failures do not really remain so. Unexpected results or the new, but unexpected, synthetic product is of interest. The researcher who is ready and able to recognize this situation then has serendipity. Open-mindedness and observation are keys, though. Many others would just see results different than expected and label it failure. The scientist who is looking for opportunities might ponder about the difference between expected and observed results, then moving on to alternate courses for the further research. Serendipity has both an observant experimenter and one who is willing to change course if an interesting, but unexpected, discovery does happen. Some examples are given in the chapter "Curiosity and Wonder".

It is human nature to be disappointed in these failures. Enthusiasm for what seems to be an innovative breakthrough idea is hard to give up in the harsh reality that nature throws on it by making it not work. Accepting failures as part of doing experiments does not mean not being disappointed; it means learning from the failures, thinking again, and trying other ideas. Another aspect of human nature is that these disappointments soon pass, especially if an eventual success follows through by finding an alternative approach. The successes are what set the overall tones of someone's research. These are what are chronicled in the literature, are esteemed by other scientists, and that leave a legacy.

Any young scientist who becomes distracted from this

further work because of an emotional or psychological feeling of personal failure needs to be helped through it by being told that experimental science works with both failures and successes. If you never fail, you never have risked. If you never have risked, you never have explored the unexplored.

2.2 Keeping Current – Always Learning

Many chemists, when starting out their careers in industry or government laboratories, think that their education has been completed. (I exclude academia from this statement because the attitude towards continuous learning by professors is much more open.) This is far from the reality of maintaining a dynamic career. New material continuously arises, even in a field in which someone is an expert. I, as an example, wrote a comprehensive book on the chemistry and analysis of the larger polycyclic aromatic hydrocarbons only a few years ago. At that time I was very immersed in the knowledge of these compounds. That, however, did not mean I would stay so without the effort of keeping up with new discoveries. It has to be continuous, even in the periods in which the book was being written and published.

This type of learning is even more critical when new topical areas and new techniques arise. Becoming expert in them must be done before someone can do quality research in that area. Reading the latest journal articles is not sufficient. The fundamental basics must also be learned from review articles and books. If not, your foray into this new area may overlook key factors and seem amateurish to others working in the area. For example, I delved into fluorescence spectroscopy in some of my research, but had to first learn about molecular energy states, the optics, energy losses through quenching and molecular collisions, and other topics. Then I could reasonably do the quality of laboratory work to write a manuscript or present a talk that would be accepted by those already doing fluorescence research.

This self-maintained and structured learning is a minimum in one's continuing education. It both maintains your expertise in your current fields and sets up diversifying into others. Formal education is another aspect that every scientist should consider.

Although young scientists consciously know that they will continue to learn throughout the careers ahead of them, they do not often foresee taking classes or other structured learning in those futures. In reality, most companies and government agencies have programs supporting continuing education. Many companies even have continuing education as an element in performance reviews and career planning. They provide flexible schedules to accommodate the taking of classes and reimbursement funds to cover their cost. Thus, formal studies do not have to end upon the finishing of graduate school. They only become less intense, less frequent, and altered in their context and content.

Douglas Lane, an atmospheric chemist with Environment Canada, passed on one of his favorite quotes. The British author Niran Chaudrey said "Every man should know everything about something and something about everything." This is a wonderful idea. In order to be successful, a scientist's curiosity and interest must continually be honed. The chances for innovations increase with broader knowledge as concepts from different fields are applied to each other. Interactions with researchers in other fields can lead to collaborations or at a minimum to very stimulating discussions that are from different perspectives. On a more personal note, learning from many areas keeps the thinking processes sharper.

Dr. Lane went on to describe his interests in astronomy, geology, anthropology, archeology, marine biology, and various types of history. He ascribes his ability to communicate with a variety of people to this diversity of knowledge. Being this diverse in interests is not unusual among research chemists. It is the norm. Inquisitive minds that can grasp complex concepts of science do not turn off once they leave the laboratory. Fostering other interests will increase your mind's ability to think and absorb new knowledge and provide you with personal interests. Occasionally you will find that certain facts you know enter into your science or are a common interest with a colleague.

22

There are many options in how chemists can continue their education after completing their degree work. There are formal ways, such as taking courses at nearby colleges and universities, taking short courses, taking correspondence or online courses, and taking internal courses offered by one's company or institution. Informal learning includes reading the experimental literature, review articles, and books.

Many societies offer short courses in specific technical areas, either in tandem with society meetings or alone in a series of venues. These are an excellent way to gain in-depth knowledge or learn the basics of a new technical area because the instructors are leaders in research in the topical area. The intensive approach of short courses is often enough to get someone started on working in a new technical area. At a minimum these courses offer a resource of information or direction as the instructors are both accomplished in the research topic and familiar with answering questions in the topic.

The taking of classes is also one way to explore new career directions without having to change one's job first. If someone thinks that he or she might like shifting into management, a financial role, work in human resources, or other types of work other than bench research, then taking a class or two in that area will expose the person to the issues in those types of work. Thus, a person thinking of moving into management might take courses in project management, creating a budget or business plan, or how to handle a diverse team. Another advantage in doing this is that the others in the class and the instructors can give further perspectives and insight in working in the new field or in changing careers and its difficulties and rewards.

Another, less thought of, way of learning through classes is in the teaching of one. Being the instructor or professor of a college or graduate school level class forces one to learn thoroughly a subject area. It is something that can be done by anyone in an industrial or governmental position if they wish to pursue it. This often involves learning aspects or details not learned in one's own taking of the class in college or graduate school. A few years ago I taught a class in experimental statistics, how and why to use the different statistical tests and approaches. In preparing for each of the lectures, I looked for hours through several texts. The varying descriptions and developments of the mathematical bases for each statistical treatment clarified my own understanding of each. This was absolutely necessary since I had to answer the students' questions, but also helped me in my own use of them. An alternative that can also be challenging is teaching some of those short courses described earlier.

If an individual does not have the inclination or time to take a full course at a local university or does not have good opportunities to take the appropriate short courses, then all is not lost on continued learning. One can take the time to regularly scan Current Contents or other similar sources of the listings of journal articles. Some of these are available as online or through e-mailed subscription, making them accessible throughout the world. One can scan the table of contents of any readily available journals, particularly keeping an eye out for review articles.

A person who queries all of her or his colleagues, will find that a surprising number subscribe individually to one, two, or a few journals. By gathering "who reads what", she or he can keep abreast of articles in many of the leading journals without either increasing one's own subscriptions or relying on the slow routing of those from the technical library. Increasing one's intellectual network can be very beneficial. Others who know your research interests may spot a useful article in a journal that you might not even be aware of. With the huge number of diverse journals being published today, this is not unlikely.

Keeping an eye out for review articles must be a regular part of looking through the literature. Although most journals do not publish many review articles, they tend to highlight those that they do. This makes scanning their contents easier. Certain journals, such as CRC Critical Reviews, or books, such as the plethora of annual reviews in various disciplines or topical areas, are a must to look at. Contacting one's technical library to ensure this is a task well worth the effort. The annual research review issue of Analytical Chemistry, which appears as the June 15th issue, is a very good overview of analytical chemistry topics. The focus of these alternates between even-numbered year issues containing reviews based on analytical areas such as electrochemistry, gas chromatography, atomic spectroscopy, and chemical sensors and odd-numbered year issues focused on application areas such as polymer analysis, crude oil and its products, ceramics, or pharmaceuticals.

Many companies support continual learning by having technical libraries or subscribing to journals that are routed throughout the interested scientists. This can be a slow process, however, since the flow stops on any individual's desk who is not a diligent and regular reader. Structuring one's time to timely read the literature by stopping in the technical library, keeping the flow of routed journals fast, and sharing one's own journals all help. If one does this, networking may also be an aid. As others read journals, they will pass on an interesting article. If everyone has their eyes out for topics that are of interest not only to themselves but to others, much more of the wide diversity of journals can be covered.

Many people think it is a huge barrier to sieve through the many journals. Some professors even do not regularly read journals, but rely on their graduate students to do this and gather interesting papers to the attention of everyone in the research group. It is a sad consequence of there now being both many more journals and their covering a wider breadth of topics. Others, however, have not personally succumbed. They still regularly look at a limited number of journals key to their work and scan tables of contents for others as time permits.

The biggest barrier to continued learning, especially of the informal type such as reading journals, reviews, and books, is not a lack of resources. It is most often a lack of prioritizing and scheduling. A scientist must set aside the time to keep current and to learn new areas. This should not be defined in vague terms such as "I plan to read for two hours a week".

Although setting a target is a first step, unless the person also includes specific time slots for this purpose, the aim quickly becomes optional. It gets precluded by other priorities, things that pop up and will "Only put it off this week". If a person starts to be more flexible in scheduling this reading time then it becomes marginal and optional. Finally it will be rarely done. Setting aside several regularly available time slots keeps a person diligent. If the schedule needs flexibility, then having more time slots than needed is one solution. A person can use some to read the target amount of time and others to handle crises as the needs arise. If in one week less time than planned is spent on literature reading or learning then the next week's time must be targeted to be greater.

Overall, continued learning must be a priority or the scientist faces obsolescence and a career of more mundane aims rather than groundbreaking work. Science is forever progressing and without determination and diligence, it is easy to be passed by.

2.3   Specialist or Generalist?

One of the current trends in the various disciplines of chemistry is the strong emphasis in research to be very tightly focused in certain applications areas. For example, in analytical chemistry there are the fields of proteomics or recombinant DNA analysis or characterization of nanostructures. Scientists from many other fields have converged to help create these new, very hot, topical areas.

Each of the resulting hybrid subdisciplines of study can be very dynamic. The recognition, number of publications, and levels of funding all rank high within any of these hot topical fields. The keen interest draws both experienced scientists from a wide variety of areas and graduate and post-doctoral students; the latter group because they wish to work in research groups creating such exciting, high-profile discoveries. This, especially, is a major positive aspect since many of the brighter, more-capable students in the past several years have gone into computer science, electrical engineering, materials science, and other fields that have driven the development of the Internet and our cyber economy and society.

Are there any negatives to having such keen interest in these narrowly focused specialties? The answer is yes. The most obvious example is the corresponding move away from a generalist and more fundamental approach to the discipline, in this case of analytical chemistry. This shift also includes a move from any of the fundamental specialties of the discipline when they are not delving much into these hot topics.

Some may ask if fundamental and general knowledge of a chemistry discipline is even needed. The answer to this question also has to be yes. This is so if there are needs for the scientist to understand the reasons why things are done a certain way rather than just doing them that way.

Each scientist must balance this trend to specialization with enough general knowledge to make innovation, versatility, and collaboration possible. The thought should always be that specialization is focused on specific research topics and projects, but in order to do these well there must be fundamental foundation for all of the work. Good work in any specific area must be in line with the fundamentals of experimental science. For example, is the work reproducible?

Being a specialist does not mean that the basics in that field are any less. In fact it should mean an even higher standard. A person who becomes a specialist is still judged by the others working in that area. Although some may not be totally versed in the basics that underlie the work, many others will be. They will look at work done without regard to accepted practice as shoddy unless the alternative way is a provable improvement. For example, someone might use liquid chromatography with mass spectral detection, relying on a mass spectral library match for peak identification. If the results show identifications that are inconsistent with the liquid chromatography retention, then others will think the work is unprofessional. An expert in LC-MS would look for consistency in the two joined techniques.

In organic chemistry some of these fundamentals are the mechanisms of reactions, the effects of temperature, reactants, catalysts, and the medium, and the influence of other impurities, the difficulties in identification and isolation. In analytical chemistry some of these fundamentals are the valid use of statistics and sampling methods, the competing interactions of stationary and mobile phases, and the interferences of matrices and impurities.

An example of the good use of fundamental knowledge can be drawn from the current use of combinations of chemical analyses, the so-called hyphenated techniques. The best possible analytical methods are developed for an optimum mix of speed, sensitivity, and selectivity; if full advantage of the power of combinations of methods into the plethora of hyphenated techniques is to be realized.

The optimization of hyphenated methods is a good example of the need for general and fundamental knowledge. Gas chromatography with mass spectral detection (GC-MSD) is a commonly used technique. How often do we find a GC-MSD analysis of less than adequate quality where the chromatography is only looked at as a sample introduction system for the MS? The selectivity of GC phases, gas flows, the shape and range of the temperature gradient, and other obvious variables are not often examined by someone with a strong mass spectroscopy background. Conversely, how often is the mass spectrometer looked at as only another type of GC detector? This happens when the researcher is a chromatographer. Any fragmentation patterns or the use of chemical ionization for selective detection of specific compound classes are rarely used by someone with a focused chromatography background. Thus, from one end the combined technique is optimized for the chromatography with little sophisticated use of the mass spectrometer. From the other end, the mass spectrometry may be very elegant, but the chromatography may be overly simplistic and inefficiently used.

Organic and inorganic chemistry are also prone to a loss of fundamentals. Many scientists think of these as cookbook chemistry, sets of simple and easy to copy recipes. New syntheses are created in this mode by changing reactants or catalysts to see what new products are made. The approach is not fundamental and directed by the basic understanding of the reactions. It is only a matter of mixing many permutations and hoping that one will result in a novel compound that meets whatever use the research is aimed at.

Two scenarios occur which contribute to the specialist approach becoming too prevalent. First, for students the periods of course work and graduate school research are learning periods of limited duration and scope. You can only cram so many courses into the required curriculum. If a student takes more involved biochemistry courses in order to do proteomics, what other courses must be left out? Basic courses in instrumentation, statistics, and the fundamentals of separation science or spectroscopy are often the casualties. They are not as directly applicable to the student's research goals. They become ever more optional.

The second scenario occurs for the experienced researcher. With so many rapid developments to keep up with, a researcher in these hot areas can spend all available learning time just on the focused field being worked on. If this researcher also has to learn this hot topic while being expected to develop new things in that area, then there is usually only an on-the-job training that can go on. This tends to be focused on only the practice of the research, how to do it only in the ways that are immediately needed. The fundamental whys and what-fors often take too much time and effort to learn. This often means that the true power in the variations in methods are not learned and realized.

This scrambling just to keep up can be very difficult, especially since this period in most careers also coincides with many things developing in one's non-scientific life. The balancing act is difficult. In order to remain a versatile, viable researcher, however, there must be some time set aside to read the literature and to learn the fundamentals. Some of this time should be spent learning new things and getting a better grasp of the basics of analytical chemistry and its various techniques.

There was a classic example of this in my own work experience that highlights the fallacy of becoming too narrow in technical knowledge. A person had the problem that he needed to know the amount of unreacted alkene in a polymer. He knew very well the NMR spectroscopist in the group I was in. They talked over this problem and the solution that was used was NMR rather than a more suitable technique like gas or supercritical fluid chromatography with FID or mass spectral detection. The latter techniques were readily available in an adjoining laboratory. The NMR spectroscopist was so focused on his own work that he did not know much of things going on nearby or understand their value.

It is the exceptional researcher who both recognizes this dilemma and can set aside the time and extra effort to build up fundamental and diverse skills. With the flood of published papers in a hot topic, it is difficult enough to even keep up in the growing plethora of established and new journals publishing in that area. But if one is to do the best that one can, then efforts must be made to broaden one's knowledge and to more fully use the techniques and methods that are available. An occasional broader reading of journals, discussions with scientists in other fields, and sitting in on sessions covering other areas while at conferences can all help.

Some might ask "Is there even value to a fundamental or generalist background?" The answer must be a loud and unequivocal Yes! Without a fundamental understanding of the techniques used, method development does become more trial and error rather than educated guesses at probable solutions. The chances for overlooking a simple, key, but fundamental fact also arise.

When I was in graduate school, all analytical chemistry students had to take a course in experimental statistics. It was taught by a professor who emphasized the fundamental mathematics behind the simple statistical approaches. He was extremely rigorous because he believed that as professional scientists everyone should use the tools correctly. I, like many of the students, just saw this course as an unnecessary impediment on our road to our degrees. It was not until years later, when I taught a similar, though less rigorous, course that I finally understood the concepts like outlying data and how to validly identify and reject it. I looked at my older laboratory work and saw that I had not properly applied the rules of statistics. I finally realized that the understanding of fundaments is the basis for the best work possible. With this I finally appreciated that professor's emphasis on the basic understanding.

There is a need for every chemist to balance the specialization driven by today's research with a more basic knowledge. The generalist approach creates a framework in which individual forays into specialties can easily occur. With a wider ranging background that is strong in the basics of many fields, one can work on specific areas within specialties such as syntheses, analyses, physical properties, and many much more specific areas.

The fields of chemistry are becoming more wide ranging and overlapping. Are they, at the same time, also remaining somewhat unified or are they becoming more fragmented and disconnected? Is our work becoming more diverse and focused on specialized topics while maintaining the highest qualities? Or is it losing its depth of understanding and becoming only passable science for the sake of discovery?

One of the strengths of chemistry has been its connection to many other fields through application on commonly shared techniques and its role of applying science directly to use by society. The linchpin for this has been that chemists help the other sciences that use our techniques to understand the work and to ensure that it is done properly. This role of being the keepers of fundamental understanding of chemistry must remain as a key one.

2.4 Being a Part of the Scientific Community

2.4.1   General Remarks

One of my favorite quotations is by the writer and metaphysicist John Donne in his "Devotions upon Emergent Occasions: Meditation XVII" (1624). "No man is an island, entire in itself. Every man is a piece of the continent, a part of the main." This refers to the connections between every person and society. Every scientist is analogously a part of the scientific community; a society that connects and supports every scientist.

There is a large community of research science. There are many formal organizations that help give the scientific community structure and connections. These are the professional societies that exist in most nations and the topical societies focused on specific research areas. Both types serve different purposes and advantages for the researcher. They augment the informal structures that each researcher creates in networking.

One of its most common aspects is the dissemination and sharing of ideas and new discoveries. This takes place through two main media. The professional societies play a key role in fostering some of both. The first is those gatherings of numerous scientists at conferences, colloquia, seminars, and other meetings. The format is often spoken presentation, but includes poster sessions. The second medium is publication in journals, books, and other printed or electronic forms.

With a wide variety of general and specialized societies, conferences, and journals, it is difficult to choose among them given the limited time and funds that can be used to be active in societies and conferences. As a corollary to this, choosing journals and books in which to read and write for is also difficult. Choosing can help in the other aspects that influence success, such as networking, collaborating, and staying creative and diverse.

2.4.2 Societies – Is it Advantageous to Belong to Professional Societies?

The large meetings of the national societies are a broad-based forum where sessions cover all disciplines. The chances of being exposed to new topics are large. Collaboration along cross-disciplinary lines can be created after hearing presentations or directly talking with interested chemists working in other areas. These can also be a good forum in which to present at if the speaking skills need to be honed. There are not the pressures that you may feel when speaking to a group who are all experts in your field.

The national professional societies are beneficial in other areas than just the technical ones. I will speak from my own experience as a member of the American Chemical Society (ACS), but the other national societies offer many of these same benefits. The ACS is involved in many activities that relate to career issues and offers resources to members. These include a usually large number of advertisements for job openings in its weekly magazine Chemical and Engineering News. There is a corresponding searchable database accessible through the ACS website. This also allows job seekers to post resumes and be notified by electronic mail of a possible job match. There is also an ACS network of chemists who serve as career councilors for those seeking jobs or thinking of a career change. The ACS also offers a variety of coverage for professional liability, life, disability, and medical insurance.

These are some of the direct benefits of ACS membership. There are also indirect benefits. Large societies can track
employment trends and conduct surveys of salaries and in these
more uncertain times for job security, surveys on job termination
variables such as severance pay, the length of forewarning time,
and outplacement services.

One should balance membership and participation
among the larger societies, such as one's national society, and the
small topical ones. The smaller societies sponsor journals and
conferences that are analogously smaller in focus. The members
in these are all working in areas that are similar to yours, so
fruitful interactions are common. The chances of collaboration
are high. Involvement in a small society greatly increases a
person's visibility among her or his peers. This is the old adage of
big fish in small ponds. If you serve in any capacity in a small
society, a good proportion of the members know of it.

2.4.3   Being Involved in Societies

If one is so inclined, it is possible to become involved in the
operating of a society. For the large national ones, this may start
as involvement in a local chapter or section in the organization
or in one of the more specialized divisions. There are always
needs for help. Many more people utilize a society's benefits and
services than belong to it. Many more belong to a society than
the numbers who make the society work.

For the smaller societies, usually only a small core of
the members is actively involved in making the society operational. This gives more opportunities, but more work to be done.
Sustaining membership and vibrancy in small societies is an on-
going task. A career change in a few key persons, such as
retirement or a change in research interests, can drastically affect
the dynamics that run a small society. An enthusiastic person can
have a major impact in the visibility and membership.

Conversely, a person who volunteers to do things and
does not can be devastating. These are particularly true for people
who become officers of a small society. Since the society is small
and only certain members want to be involved in the operation of
it, there is a constant need for new people in the running of the
society. Without that, the organization soon seems to be a small

clique, rather than a group of interested researchers. Also, with so many demands on so few, the active members may become overwhelmed or jaded and join the ranks of the uninvolved.

Becoming involved in local sections, society subdivisions, or small societies starts with attending the meetings. At them you learn what goes on and who the active members are. After you become comfortable, which can even happen at your first meeting, then you can approach the active members and volunteer to do whatever you could do that matches their needs. If you ask, the number of active and involved members in any society is a surprisingly small fraction of the membership. This gives lots of opportunities to do both needed and significant work.

After more time and involvement, you may even be willing to serve as an officer. This means more commitment in time and energy, but also gives more visibility and involvement in more important issues. Your other interests, such as in any publication the society may be involved in, can join with your involvement in the society. Most national societies and many small ones have committees for publications, conferences, awards, training and career issues, and many other areas.

One type of involvement in committees for societies is not by self-nomination. Most societies give awards and large ones such as the ACS give several. Behind each of the award announcements lies a hidden committee or jury that reviews the nominations and selects the winner. The people on these selection committees are chosen because of their technical expertise in the area of the award. Membership is voluntary and generally recognized scrumptiously in order to keep the process confidential. I have served on a few of these and I can say that the nomination, review, and selection process is interesting to be involved in.

2.4.4   Conferences – Is it Advantageous to Attend and Present at Conferences?

One of the biggest benefits can be said in one word, "Networking". Making connections with others is advantageous in doing better research through exchanges of ideas and results, setting up collaborations, and in the non-technical areas of job searching and involvement in the profession of science. Attending conferences and symposia make a person more visible and allow for direct, personal contacts. These give a much better impression than being published, corresponding, or talking in telephone calls. This direct and personal linking is many times more effective than those others. This is human nature. Knowing a person, having an image of her or him in one's mind is a strong link.

The smaller conferences and symposia devoted to specific areas of research are quite different than the large national meetings. These meetings of a few hundreds of attendees are very focused on specific research areas. The level of expertise and familiarity to the research are much higher. This leads to very dynamic interactions where ideas for new research can arise both from the presentations and the informal discussion and where collaborations arise because of complimenting knowledge, equipment, techniques, chemicals and other resources and give mutual benefits.

The amount of networking that can go on at a small, focused conference is doubly high because of the similar research backgrounds of the participants and because of the smaller size. The moods at such meetings are more relaxed and the small number of people naturally lends itself to more interactions. With few concurrent sessions, the participants do not rush from session to session. Often these smaller meetings contain breakfast, lunch, dinners, and social functions such as mixers and tours as structures in the program, increasing the networking opportunities.

Some meetings, such as the Gordon Conferences in the United States, are specifically designed to create this intimacy for idea exchange, debate, and potential collaboration. These meetings are often located in a scenic, but very rural area or in a small out-of-the-way town that has a guest hotel or university. There are usually many social events, mixers and meals designed to increase everyone meeting and interacting. The atmosphere in these meetings is much more relaxed and casual, including such activities as hiking, fishing, or mushroom hunting. Some of these meetings are open to any interested scientist, but some like the Gordon Conferences are by invitation.

In contrast, large meetings can be so full of hustle and bustle, with everyone going from one session to another, that it is difficult to meet new people. This usually happens in these large venues only as a minor happenstance during coffee breaks (and these are seldom directed and fruitful) or through introductions after presentations. In a large meeting, the focus is more on knowledge gathering, listening to the presentations for new and interesting work. Several researchers said that this was the most efficient source of information since much can be gathered in a few days on listening. The information output at a conference is very concentrated.

Certain meetings, such as those of the Pittsburgh Conference in the area of analytical chemistry, are hybrids. They are large and multi-topic, but the focus in each symposium or session can be very focused. Its largeness, with well over 20,000 attendees for every one of the past couple of decades, means that the exhibition portion is an attraction in itself.

As far as presenting a research talk at a conference, there are several advantages. One obviously is that you become better known to others in the field. Interacting directly with someone, even if it is only in seeing them give a presentation, makes much more impact than reading that person's name on a publication. Actually talking to them is better and getting into a detailed discussion of mutual research interests can lead to collaboration.

At a certain point in a career, the possibility of chairing a session at a conference will arise. Alternatively, one's experience might lend itself to putting together a session. These are great opportunities to gain stature, network, and learn. If you put together a focused topical session at a meeting, you can ask many of those who are doing the better research in that area if they would like to participate. If they cannot or choose not to parti-cipate, then asking them still creates a positive impression. If they do participate, then talking to them involves interacting before-hand, during and afterwards to arrange symposium session publication. The chair makes sure of the available facilities for visual format (slides, viewgraphs, computer projection, etc.) and informs the speakers so that they can be prepared. As far as the session itself, the chair must make sure every speaker is present and gathers some introductory information. As with chairing any session, the chair keeps the speakers to a schedule, keeps the question section to its allotted time by asking those with more questions than allowed to ask the speaker during a break, and must be prepared to ask questions in the event that no one in the audience does. Since this was an invited session, the chair should especially take the time to thank the speakers.

Another role that can happen is in helping assemble a technical program. This is much more than just scheduling talks. The topics offered by those submitting presentations to a meeting are diverse. A committee usually divides these into topical areas and decides if a poster or oral presentation is preferable. In some cases, specific speakers must be invited to participate in a topic. Finding people to chair the sessions is also a task.

2.4.5   *Journals and other Publications – Is it Advantageous to Publish Papers?*

To the academician this is a rather silly question because many things, the most important being tenure decisions, are assessed at least partially by one's publications. Professorships most commonly involve a mixture of teaching and research. The quality of the research is in-part measured by the number of publications and into which journals they are.

The need to publish is particularly true for scientists in industry. It is not always apparent in a particular position, but when a career's goals are looked at, it is much needed. As opposed to academia and governmental positions where publication is highly encouraged or even demanded, industry does not have any inherent affinity towards publishing. The lone exception may be in patents.

In today's much less secure job climate, however, this can result in the scientist having little work available as a ready resource in a job search. In certain fields much of the research is confidential and proprietary. Discussing it with another company, even as part of a job interview, can be legally hazardous and ethically questionable. If the scientist, however, has made an effort towards publishing, then she or he has something in hand.

This may seem to be a paradox since the work is proprietary, but published work can be planned as a contingency. For example, general aspects of a new synthesis would not necessarily be a trade secret. Neither would a report on development of a new method that utilizes model compounds. In some cases this may be through experiments on some generic

38

material such as a standard reference material. Getting published in this fashion can be done in many areas of research.

There is a choice of which journal to submit to. This mainly depends on the subject matter. The best clue for choosing is looking at the articles published in each journal. Do certain journals publish more than others in the topical area? Is the tone of research fundamental or applied? If the research has cross-disciplinary aspects or potential uses in other areas, is the journal broad based? If the work is important, what is the average timeframe between submission and publication, as well as is there accelerated publication such as an electronic or online version? How prestigious is the journal (in academia in particular, this can be a criterion in reviewing accomplishments for tenure)?

Typically you want to submit to an overall range of journals. This reflects both having a diverse set of research areas and a need to be seen in both topical and general area journals. This also strengthens your research as the criteria for publishing and the types of reviewers will vary from journal to journal. More critical input can help you see areas that need more emphasis.

Publishing also accompanies some of the other aims described in this book. Through publishing, opportunities for collaboration can arise. Others read of your work. They think of ideas that cover additional areas or combine elements of their own work in combination and compliment to yours.

Writing review articles in a cross-disciplinary fashion exposes others to the basic precepts of your own areas of expertise. This also creates potential collaborations. Review articles also are a very good way for graduate students and others in the beginning phase of their careers to become better known. Writing a review relies most on a thorough literature search and then organizing the results in a concise and readable form. It does not take months and months of innovative research and it does get read by many people. Writing reviews would be one starting point in networking.

The germination of a review article is not only the receipt of a request from an editor asking for one on a topic of the journal's interest. Many are started when a researcher realizes that a topical area is heating up; attracting more interest. A review of the fundamentals and history of this research area, as well as its potential applications, may then be written and submitted. Another need for a review article is shown when the file for a topical area in a personal library (see the description of this in The Toolbox chapter), there has been no other reviews of the topic in book chapters or journal articles for several years and there has been numerous articles from a wide range of journals since the last review. If an editor thinks that there is sufficient interest in the journal's readership, then these reviews will be published.

The publishing analogy to chairing a session at a conference is to be a guest editor for a journal and put together a special topical issue. This has the same advantages of allowing you to interact with other leading researchers at the different points in production. Most young researchers do not know that many journals are open to special issues or sections of an issue that focus on one research area. If you have a good idea for a topic, it can easily be written up as a proposed special issue. First think of a few colleagues or other important scientists who might contribute. Pass the idea by them; see what they think of the idea and if they might contribute. If the response is positive you can then send your proposal to the editor. Proposals of this sort can do little harm. Even if they are rejected, you have created stronger linkages in your network of colleagues by asking them for feedback and participation.

2.4.6   Reviewing for Journals

One role that any scientist can take up in the publishing process, in addition to writing for publication, is to serve as a reviewer. How does a young scientist become a reviewer? There are many routes, but one most do not think of or do is self-nomination. It seems too presumptuous and too audacious. In reality, journal editors often scramble to find technically qualified reviewers for submitted manuscripts. The reviewer must have expertise enough to understand the content of the manuscript and to judge its merits and innovation. Contacting an editor to offer one's help is quite alright. Another route is to offer that role to others in the field so that they may list you as a proposed reviewer.

Being a reviewer has one tangible benefit to one's own research. The new research described in the submitted manuscripts is seen earlier than its publication. It also forces each reviewer to more thoroughly understand the background and basics in the field. This important role cannot be taken lightly. Checking references and making sure that any other relevant work is cited and assessed requires some work. Thinking over alternative conclusions is another area that takes some thought.

The most difficult aspects of reviewing, besides the time committed to doing it, is giving negative feedback when it is needed. This is not easy in any context, but as a reviewer it is sometimes even less easy. If work is flawed or the manuscript is poorly written, then the reviewer must either reject it or send it back with major revisions. It does take the right attitudes to do this, which include enough confidence in one's own technical understanding and assessing and also being able to overcome the empathy with the authors. It is natural for a reviewer to think how she or he might react in getting such a review. Receiving a rejection is always bothersome, but if the work is truly not up to the caliber of the journal it should be rejected. It will also be a bad reflection on the authors if it is published as is.

2.4.7 Advisory Boards and Editorships

When the research career has grown so that there is some strong degree of awareness of a person's research, she or he may be asked to serve on an editorial advisory board or even as a technical editor. I cannot go into what leads to these offers other than to say publishing, networking, and involvement in the journal as a reviewer all can be factors that bring your name to the forefront when nominations are asked for. What does being an advisory board member involve?

Editorial advisory boards are the sources of technical guidance for a journal. They often tell the publishing staff what areas are highly active and deserve a review article. They are asked who might be approached to write those reviews. They also solicit submissions to the journal from their colleagues and networks and are strongly encouraged to submit some of their own work to the journals in which they are advisory board members. Much of this is done through correspondence in one form or another. Actual physical meetings of an advisory board are rare. These are often coordinated to be held during the larger society meetings and large specialized ones. Attendance is only encouraged and not required. If you do attend, the discussion can be lively and very informative on emerging research areas. These also are another opportunity to network with eminent scientists.

Technical editors have many more responsibilities in addition to those that advisory board members have. They generally are the ones who offer candidates for the advisory boards. They decide on the topics and solicit guest editors for special and topical issues, as another example.

Their most important role is in the manuscript submission process. The technical editors receive the submitted manuscripts either directly or from the publishing staff. They must look through the manuscripts and assess what technical area these are in. They must then choose appropriately qualified reviewers, so a database of names and contact information must be on hand. They then must send copies of these and review forms to the required number of reviewers. They must ensure timely reviews and a consensus response to the submitting authors. If one of this first set of reviewers cannot be timely or has a conflict of interest, then an alternative reviewer must be selected. The technical editors look over the revised manuscript and decide if reviewers need to assess it again. The final consensus can involve reconciling different opinions from reviewers or between the reviewers and authors, either personally if the expertise is theirs or calling in another person as a referee. Upon acceptance, they inform the authors and send the manuscript to the publishing staff for preparation of printed proofs.

For some journals this process may involve tracking dozens of manuscripts through this process simultaneously. Although many journals now do these processes electronically and online through e-mail and the Internet, the steps and responsibility that all of the steps are done in an acceptable time still belong to the editors.

2.5　Thinking – Curiosity and Wonder

The other half of the thinking that leads to creativity in discovery is the curiosity to learn new things and to find out what has not been known before. It pushes knowledge forward and balance the skepticism. One personal reward for this is the feelings of wonderment and even awe at times.

The inherent curiosity and the wondrous joys of finding new things is a fundamental key part of scientific thinking. Throughout my career I have talked to many scientists and found one striking similarity in our lives. Most of us first became interested in science as young children and had decided on a career in science at a much younger age than those people engaged in other professions.

Many, including me, knew we wanted to be scientists well before we reached our teens. My education and career choices were only fine-tuning this decision into more specifics, until I finally chose analytical chemistry as my graduate school focus over ten years later. I do not think that this similar youthful decision to become scientists is a coincidence. The joy of discovery and the curiosity that emerged in childhood can find full expression in the profession of research scientist.

One recurring theme in the discussions with other successful researcher chemists was the enjoyment of the work. Many scientists emphasize that it is fun to do experiments in the laboratory. They referred to experimenting as "playing in the lab". The childhood feelings of discovery are still strong in them.

The curiosity of how things are; what makes them behave the way they do; is a key part of the researcher's mind. The exploration into this eventually leads into discoveries. These often contain what I call the "*Wow factor*", the joy at finding out something novel that explains why things are as they are. This factor is increased when the discovery encompasses explanations that then lead into further questions and answering discoveries.

What does this have to do with successful research? I think that in order to sustain a career in research, an individual must remain enthusiastic. She or he must have a hunger and curiosity for discovery and a joy for when it happens. Constantly learning and delving into new areas feeds this hunger. This hunger must both be constantly satisfied and subsequently grow back into a need for even more new science. Sustaining this hunger, the curiosity for new knowledge and discovery, is not inherent. Many things can be done to make it a natural part of one's career.

What things help maintain that enthusiasm? What sustains the curiosity and wonder? How does a scientist still keep that *Wow factor* high?

I have touched on some of these factors in other chapters. One is exploring new technical areas, diversifying one's research focus. In order to diversify, a scientist must first learn to be proficient in that new area. This takes work and study. Another factor is sharing and collaborative behavior and thinking. The discussion of ideas gives different perspectives, brings in knowledge that the individual was unaware of, and pushes creative thinking....doing novel things.

Certain good habits also help set one up for continuing discoveries. We all are busy, so keeping up with emerging science gets difficult. One major thing that must be done is structuring one's schedule for time to search and read the literature, to correspond with other scientists, to write papers, to attend conferences, and even to think about experiments.

As an example, some scientists go to conferences with their schedules mapped out, planning which presentations to see.

The focus on attending talks can be so strong that an individual rushes from room to room, catching as many in the day as possible. This is one way of gathering knowledge of what is going on, but not the only one, I also think that it is not the most valuable way. Being more selective in which talks to attend, gives a scientist time to listen more closely rather than watching the clock. The times between presentations, particularly after a session, become opportunities to ask more in-depth questions to the speakers, to exchange business cards, to discuss what is going on at the moment (since most meetings have submission deadlines months before), and to connect with the other scientists. This is allowing for the time to build a better network and to look for collaborative opportunities. There is a strong mental stimulation when you interact with others who share your curiosity and who can think creatively.

These exchanges can contain many nuggets of *Wow-factor* science. I once participated in an impromptu discussion with three other scientists in a hallway during the coffee break between sessions of a conference. The discussion was so dynamic and creative that at least a half dozen research papers resulted from the ideas discussed. Afterwards, that ten-minute break was much more important to each of us than many of the two-hour sessions.

Another good habit that is connected to this is to not stand still. I have touched on this in the chapter on diversifying your interests. Working in a new area of research requires learning what has and has not been done in that area. The limitations can be translated into opportunities by applying new ideas and techniques (even if they are only new in that area, but are established ones in another). A corollary to this is always "How would I have done that?" when one reads a publication or hears a presentation. Another good habit or attitude is to not get caught up in topical or disciplinary labels, not letting these become boundaries.

Always asking questions in one's mind, being curious and wondering on the nature of things, are part of maintaining a vibrant career. This drives the learning of facts new to the individual and also sometimes results in new research in an unexplored area. Researchers must avoid assuming that something must have already been studied and understood until they are certain by looking into the previous work in that area. All too often someone thinks of a "Why?" question and then glosses over it thinking that someone must have already studied it.

Another opportunity growing out of this questioning attitude is looking at an old problem and explanation with new techniques and approaches. New insights may occur. In one of my research projects, I repeated a synthesis performed thirty years earlier. Five isomeric products were known and explained by the reaction mechanism. Using the modern combination of non-aqueous reversed-phase liquid chromatography with a full-spectrum UV absorbance detector, I found a sixth isomer. Its occurrence required a change in the ideas on the mechanism. Using new approaches and theories to look again at older research can have its discoveries and give new insights.

One type of discovery that seems to create the highest *Wow factor* is the unexpected one. Of these, the sub-category of serendipity is probably the highest in personal satisfaction. Serendipity, however, is not what many think of it. It is not a totally accidental discovery. These discoveries may arise through an unintentional manner, but once the unexpected results are found the discovery does not proceed accidentally.

There are many things that lead to an unexpected discovery. Luck is often cited as the key one, if not the only factor mentioned. Sometimes an accident may give new results, but there must be observation and insights. I think that this points out that the key factor must be an open and observing mind. Many people might run the same experiment. Most would see that the results were not what were expected and toss them aside as a failure. Only a few will see the results, assess them, and see them as a new direction of thought. This may not be readily apparent when looking at each episodic discovery by itself, but when you look at dozens and dozens of them a pattern emerges. Walter Gratzer has written a book, Eurekas and Euphorias, which recounts many of these discoveries. The initiation of the new discovery may be accidental or unexpected, but the research into finding what those meant and why they differed from the expected results is structure and logical.

As a recent example, over the past decades how many dozens of times (if not hundreds) had carbonaceous materials, graphite, charcoal, etc., been vaporized at high temperatures in a plasma or electrical arc? Yet, Kratschmer, Huffman, and Lamb[1] were the first to observe that this produced soot that was partially soluble. This led to the serendipitous discovery that fullerenes were made in macroscopic amounts. The whole field of fullerene and nanotube sciences resulted.

Finally, I will mention another type of *Wow factor* that I have felt. This is the sense of joy and pride at seeing one's work in a publication. This is accentuated when others praise the work and that publication. Over the years I have grown to feel a part of a long chain of published scientists. In reading the background for an experiment or in preparing a review, I might read of work done years ago, written not only before I was born but before my father or grandfathers were born. It awes me to think that decades from now some as yet unborn scientist will read one of my papers and use that in her or his research.

Throughout their careers, successful scientists will continually enjoy the positive feelings that result from good research work. They will feel their *Wow factors* often. They will keep curious and be struck by the wonder of discovery and insight.

Reference
[1] Kratschmer W., Lamb L., Fostiropoulos K., Huffman D. (1990) Nature 347:354–358

2.6   Thinking – Skepticism

There are two counteracting thought processes in good research. The first of these is the quest for discovery, that driving force to figure things out and find new knowledge. I discussed some aspects of this in the previous chapter on the curiosity and wonder of new things. The other type of thinking, which acts as a counterbalance, is the inertia of current knowledge and ideas that are reflected in the skeptical nature in which new results and ideas are scrutinized. This scrutiny allows only accepting those new things that can meet a certain high level of proof.

This latter thought process acts as a control on the former that ensures validity and proof before the new work is accepted. It prevents overzealous researchers from their own gullibility. Without this critical review and its need for proof, science would have many dubious and often competing ideas as its tenets. Although there was an initial keen interest in such proposals as polywater and cold fusion, this interest was driven more by skepticism than belief. These new ideas eventually did not have the proof needed and were abandoned. Conversely, ideas that were first thought of as radical when they were proposed, such as the splitting of atomic nuclei or the reactivity of the group 0 noble gases, were accepted as their proofs rose.

In this chapter I will discuss some of the aspects and thinking within these two thought processes. Although they are opposites in tone, it is through a complimentary counterbalance that these two work in synchronization.

A Healthy Dose of Skepticism

One of the major tenets in research science is that whatever is known at the moment is only a partial view. New discoveries and theories constantly refine our viewpoints. These new discoveries create the awe, the wonder – the *Wow factor* of understanding something that was unknown before. There, however, is not a wide-open acceptance of anything new. Discoveries must be proven.

In order to build truly good science, this drive for new discoveries is balanced by careful review. This starts with the experimentalist looking skeptically at the work and results. Repeatability and reproducibility, precision and variability, patterns and correlations are among the issues that are examined. The new results must meet these before there are any thoughts of interpretation.

Once these issues of quality are met, then the question "What do the results mean?" can be asked. If there are trends and correlations, the skeptical scientist looks to see what variables might cause these. Were those variables controlled enough in the experiments so that another unknown and new cause or variable might not be a factor? If the results can be explained by known theories and facts, there is then nothing novel to report. The data is only another example of the presently known theories working. The experimenter must first go through many reviews and is skeptical in criticism of her or his own work. If all of the answers point towards something novel and unexplained by current knowledge, then and only then can the experimenter start speculating on a new alternative hypothesis.

In the theoretizing stage, the skepticism must remain high. Explaining the results cannot rely on facts or laws of nature that are unprovable. If there must be new explanations, then the new scheme must have circular logic. This means if it is true to explain the new experiments, and then there must be related unperformed experiments that can be designed to verify the new explanation and its results. The classic example in chemistry of this is Mendeleev's periodic table. It explained the relationship of the elements in a pattern. In doing so, there were gaps for undiscovered elements which were later discovered. These discoveries confirmed the concept of elemental periodic families.

In recent years there have been some notorious examples of fabrication of experimental results. Two examples are the claims of discovery for elements 116 and 118 and of the electrical conductivity properties of molecules such as pentacene. In both of these cases the research was published by teams of several people, but the scientific misconduct apparently was done by only one individual in each case. If the other authors involved in these studies had exercised their scientific skepticism, requiring close examination of the data even though it came from a colleague, then the tainted work might never have even been submitted for publication.

Although those are extreme examples, each scientist should look at results with as unbiased an eye as possible. This role is mainly occupied by the reviewers and referees that journals utilize to examine submitted papers. Additionally, any interested reader has this skeptical eye on new research through the process of reading it, assessing, reconciling it with previously known work, and assimilating it into the body of science knowledge. If there was an oversight that the authors and reviewers missed, readers should point this out either directly with the authors or through a letter to the editor of the journal.

The review process for submitted manuscripts relies heavily on the knowledge and skepticism of its reviewers. The rejection rate for many journals is greater than 50 percent. The review process involves more than questioning the experiments and their results. Critical reviewing for the newness of the work (its innovation) and quality are also important criteria. All involve using one's skepticism to some degree. Is this new work or is it only a permutation of already known and published research? Is the work sufficiently repeated with all important variables defined so that the results can truly be said to be different than earlier work?

Skepticism means questioning, but still with open-mindedness that allows acceptance with sufficient amounts of the right proof. This contrasts with cynicism, which questions the new research negatively in an expectation of rejection. The standards for acceptance with cynicism are either very high or non-existent. This is also not the right mental approach to new findings. Gullibility is on one end of the spectrum and cynicism is on the other. The good researcher finds something in the middle ranges, curious to find out new things and needing proof before those are believed.

The research scientist must have defined ideas for proof. The foremost is not believability, as some think. Many scientific discoveries are very contrary to accepted ideas at the times of discovery. The noble gases were once thought to be too inert to form chemical compounds and today there are many xenon, krypton, and even neon and radon compounds. Artificial elements were once not accepted and today there are about two dozen of them. Observation of single molecules was once scoffed at and now is done by several techniques. Being skeptical or even dubious of new findings still means being able to change your opinion if proof is given.

The first proving point is reproducibility. Do the authors rely on one or a few measurements or do they rigorously repeat them? Are all variables accounted for so that the results cannot be due to another cause than that proposed? A further test occurs after publication. Can others follow the described procedure and get the same or similar results? The laws of nature are universal, so an experiment should be able to be copied.

When those are answered positively, then the questions move onto other areas that define the new results and their impact. This defines how much change in current theory must be made to encompass the new findings.

A reviewer or any reader of the published work must have criteria for acceptance, even if these are tied to conditions. For example, a new synthesis may not have the high yields reported, but it still would be a new route to the target compound. Skepticism allows for the reader to accept the reaction, but not necessarily all the other details reported. This opens the door to using it, seeing if the yields are indeed high, or even improving the method. Cynicism rejects all of it.

One of the hardest areas in which to be skeptical and neutral is when one reviews one's own results or writing. Personal bias and possessiveness naturally create a favoring of one's work. As with any other scientific work, however, it must be remembered that it all is only a partially true collection of data, ideas, and hypotheses.

No work is complete and final. All research is part of a continuum of learning by the whole scientific community.

Tomorrow's research or even today's from another researcher may increase the understanding and change perspectives. New work may even disprove what is thought of today as true and very elegant work. That is the nature of scientific research. Clinging to cherished ideas, which are especially likely when they are one's own, can keep a scientist from moving on to new discoveries. Being skeptical includes balancing the skepticism with the wonder. Too much wonder in your own work can result in you remaining behind as others see new ideas.

2.7 Diversifying

I will begin this section by giving again the quote from the British author Niran Chaudrey which Douglas Lane passed on to me. He said "Every man should know everything about something and something about everything." A widespread knowledge allows one to move easily from one research area to another. Why is this important?

One answer is that in order to stay at the leading edge of scientific discovery, one must be able to change the focal areas of research. Science discovers and the foci change. My graduate school research advisor, was a senior professor whose former undergraduate, graduate, and postdoctoral students included many well-known analytical chemists. Through guiding and mentoring these many students, he had gained a lot of experience in what it took to be a good scientist. From his many years of being a professor in the field, he often told those of us things along the line of "Do not stand pat. Technologies change. What you're doing now could be obsolete in ten years." His model for this advice was his own career, which evolved and metamorphosed from wet chemistry, through electrochemistry, through spectroscopy, through chromatography, until the final years of his career included biochemical and automated analysis.

In hindsight, his advice was exactly right. I have explored many aspects of chromatography and molecular spectroscopy and branched into several specific subtopics. Some of those areas have been outside of the discipline of analytical chemistry, including synthesis and determining physical properties. They were no less interesting than what I considered my basic areas. In fact delving into them often was much more interesting because almost everything was new. The learnings from the new areas also helped in some of the older ones.

In addition, there is a corollary to this line of thinking. The willingness to do work in other disciplinary areas is often the door to very productive and interesting research. Diversity in talents and expertise leads to diversity in applications and project areas. Being able to look at other research areas and understand what the limitations there are to progressing can lead to the opportunities of breaking through those barriers that others have not been able to.

The field I am most familiar with, analytical chemistry, can serve as an example. It encompasses two major areas, development and application. The first involves the creation of new instrumentation, techniques, methodologies, data assessment and treatment tools, and other technologies. The second involves taking these and making them of use in other areas of science. The dynamics of analytical chemistry over the past few decades has shown that much of the work that has gained the most notice in research funding and published papers has been the application of analytical technologies into areas in which they had not been used. These seldom have been efforts of the original technology developers, but have rather been by others who understand the technologies' usefulness, who also possess an understanding of the needs of the application area, and recognize that one fits into the other.

Most other areas of chemistry and the physical sciences have two or more focal areas. Organic chemistry involves syntheses and mechanisms as one focus. This is extended to include applications, natural product chemistry, and the other topical areas that entail the targeted molecules and their uses. Physical chemists focus both on the properties and behaviors of molecules and the experimental means to determine them.

Another aspect of diversity is not accepting that the disciplines of science have boundaries that divide and delineate work. My experience in analytical chemistry meant I was ideally positioned to do this since many other disciplines have a need for measurements, separation, and identification. This, however, does not mean that someone in the other disciplines cannot be diverse or branch out. An organic, inorganic, physical, biological, or other type of chemist often does this without thinking because the research does not fall clearly in one discipline. What was once thought of as hybrid work in areas such as bioinorganic chemistry or conducting polymers or synthetic medicinal chemistry are now thriving research areas. These overlapping disciplines are commonplace because investigations involve knowledge in many areas that were once thought of as separated.

From my own research history, I can give an interconnected series of examples. As analytical chemists, we rely on organic chemists to synthesize the molecules that we use as standard compounds. If no organic chemist has done this, especially so that compounds are readily available from the fine chemical sellers, then no work gets done with those compounds. In my own research area of the polycyclic aromatic hydrocarbons (PAHs), there was almost no work done on the analysis of the larger ones. Why? ... only because the compounds could not be readily purchased!

In what appeared to be coincidence, if you looked in the catalogues for fine chemicals, the number of commercially available PAHs dropped off dramatically as the carbon number increased. A quick search of the organic chemistry literature gave a few simple syntheses of some larger PAHs. The starting materials and reagents were readily and cheaply available. Once these few syntheses were performed, the rest of the work was essentially analytical chemistry. Purification involved extraction, fractionation by adsorption chromatography, isolation by HPLC, and characterization by molecular spectroscopy. A few variations on those syntheses gave more isolated PAHs. These experiments were published in an organic synthesis journal.

In touching on the first area of diversity mentioned above, I had used nonaqueous reversed-phase liquid chromatography to separate the extracted product mixtures. The separations approach could be thought of as my original base. I did separations in graduate school. With this PAH work, however, I was given and took opportunities to move into other areas. Not only did I find new PAHs in the expected situation of separating the mixtures from new syntheses, but I also found new ones in the reaction products made in replication of work done earlier by others. I identified several new isomers that had not been found in that prior work that used older separation and spectroscopic techniques.

What happened then? After the couple of initial papers on the HPLC behavior of these PAHs in non-aqueous reversed-phase HPLC, I started getting contacted by other researchers seeking samples. Some added contacts were offers to trade for my few with some that each researcher had. My dozen or so standards became twenty, then thirty, then a continuing increase of more.

These gave me the personal opportunities to expand into doing fluorescence spectrometry and the analysis for these larger PAHs in many previously unexamined sample types. Additionally, once word spread of my supply of large PAHs, collaborations in many other areas arose. I was soon receiving letters and e-mails asking if I wished to collaborate or if those large PAHs were available. Suddenly there were many papers appearing on new analytical techniques, on novel chromatography phases for selectively separating isomeric PAHs, and on large PAH occurrence and chemistry.

Am I chromatographer? Am I a spectroscopist? Am I a synthetic chemist? Am I a physical organic chemist? I have been asked all of those and have to say yes to varying degrees. Several years after I finished graduate school, I was fortunate to host my research advisor as a guest speaker where I was working then. In discussing my burgeoning PAH research, he laughingly asked if I had become an organic chemist. My path of diversity may not have been routine in its specifics, but it reflected a willingness to spread out and move into work in other disciplines. The overall pattern is of moving into new areas, becoming versed in them and in a few cases an expert, and always looking for interactions that lead towards these.

The new areas, as mentioned earlier, loop back into the older ones. My newfound synthetic knowledge led me to apply modern HPLC to several older syntheses. This resulted in the discovery of several new isomers that the older work had missed. In another example, the use of full-spectrum fluorescence detectors in HPLC was greatly aided by my earlier work on fluorescence spectroscopy. I looked at the combination as neither a spectrometer with a novel sample introduction system or as a chromatograph with a novel detector. I looked at the whole as an optimal system where strong knowledge of both parts compliments each other.

If you look at the list of publications of any eminent

chemist, you will see diversity. There might be major topical areas such as natural products synthesis or development of new separation techniques or the use of molecular modeling to assess chemical properties, but in the course of a career there will be a wide range of papers in other topics. In my case, there are some on areas such as the aqueous solubilities of PAHs, the analysis of hydrocarbons in deep-sea hydrothermal vent materials, the separations of isomeric fullerenes, and the use of perhydrocoronene as an inert spectral matrix. This variety had many benefits including exposing me to knowledge in areas that I was unaware of and collaborating and becoming friends with many scientists.

This open-mindedness is an advantage to doing research. If one looks at the natural world from a different perspective then it is evident that what scientists perceive as questions are only sets of interconnected, but yet unknown, facts. Answering these, from that point of view, has no labels and no boundaries into different sciences and disciplines. The means of discovering those facts can be open to chemists or physicists or anyone else who investigates them properly. Thus, the properties of a molecule might be found by an individual who labels herself or himself as an organic chemist or an analytical chemist or a physical chemist or some other such label.

The parochial mentality of science is not needed, but arises because groups of scientists think of work in an area of study as falling within their discipline's domain. In reality, there are no domains in nature. All of our categories of physics, chemistry, biology, geology, astronomy, etc. are human classifications. The disciplines within each overlap even more with each other.

In fact in what seems to be a real twist of fate or an irony, analytical chemists often face a prejudice from the other disciplines of chemistry, especially from certain areas of organic and physical chemistry. These attitudes are especially prevalent in some academic circles. Many organic chemists think that separations and molecular spectroscopy are simple talents that anyone should be able to do since their use of these techniques is often fairly cursory. Physical chemists involved in atomic or molecular spectroscopy look askance at the low-level of science in our applied versions of "their" techniques. Both of these attitudes are based on the belief that analytical chemists deal with a simplified world that is not real, hard-core science.

Analytical chemists must fight this narrow-minded, archaic attitude by working in these and other areas. They must show the skeptics that the in-depth knowledge of analytical chemists is valuable. This entails showing the value of both the knowledge of the theory and fundamentals of analytical methods and the understanding of how techniques can be used to answer real questions.

Conversely and in some defense of the other disciplines of chemistry, spectroscopy is not merely an analytical chemistry area. Its basis can be thought of as physical chemistry and its applications encompass organic, inorganic, environmental, and other areas of chemistry. Chromatography also is not bound to analytical chemistry. The areas of study overlap greatly. This is an advantage because capable researchers from many backgrounds can work in the same field.

The sciences must be inclusive, not exclusive. There is just too much for each individual to know and master in order to do the highest level of research. Relying on the knowledge of others and learning from them helps. Exchanging expertise, especially when working in one of those cross-disciplinary areas, can also help. There are some ways besides the personal interactions. One is by attending and giving presentations at conferences and symposia outside of your core areas. Another is publishing, particularly review articles, in journals read by scientists from other disciplines.

These attitudes do reflect the wide-ranging nature of analytical chemistry. This discipline is a core part of the work done in many other areas of chemistry and science. It is a natural flow for an individual to move from work in one area to another and another and another. By being able to do this and making the efforts to do it well and often, a research chemist insures that there is a career full of challenges and rewards.

In gaining such diversity in a career, a person might ask "Is this just moving around with the current and only being capable and not excellent at any one thing?" The old analogous saying is being "a jack of all trades, but a master of none". To me this is more a defense to resist change than a statement of truth. If you look at the careers of preeminent chemists, such as the

58

Nobel laureates or Priestley medallists, you will find changes in topical areas and diversions throughout their careers. There is a paper or two where they worked in a different area in the midst of other research. For many successful chemists this pattern is followed. There are successes in several research areas. It is not rare and, in fact, is highly likely to happen. Since all of the qualities, habits, and ways of thinking and doing things that bring success in one field of research are the same as in others, a person only changes technical focus.

2.8 Parochial Science – Possessiveness and Boundaries

How do we identify ourselves as scientists? I recently was part of a discussion where someone said "I am a mass spectroscopist, so the solution to this problem is mass spectroscopy." I paused after an initial moment to gather my thoughts for a more diplomatic reply than the acerbic and off-the-cuff one of "Fitting square pegs into round holes even if you need a hammer". I then pointed out that "That's why there are several of us involved, to get better fits so that the answer we give is valid."

I thought further on the initial comment and it strikes me as being very parochial, that is, isolated in only one part of our world of chemistry. In our increasingly specialized and focused fields of interest, we risk this narrow-minded attitude. This is brought about by many factors including limited technical understanding of other areas and the natural human psychology of "My area is better than any other".

If the purpose of science is exploration to discover new knowledge, then this parochiality is a formidable barrier. In the past two-hundred or so years, science has changed from an area where the luminaries we read of worked in many fields into one where the complexities have led to specialization. Scientists in the eighteenth and nineteenth centuries saw few boundaries and explored wherever they found interesting questions to be asked. Those who we think of today such as Antoine Lavoisier, Justus von Liebig, or Sir Humphrey Davy, thought of themselves as natural scientists and studied other areas such as geology, paleontology, and biology. There were no disciplines of organic chemistry, inorganic chemistry, and biochemistry. Others whom we think of in our modern parochial terms as physicists or biologists did work in chemistry as well. Each scientist today knows one, two, or a few areas well, but cannot be expected to work in several.

If you reverse the perspective, however, a different view arises. To think from the standpoint of facts waiting to be discovered, it is apparent that knowledge wears no labels saying "Chemistry fact", "Biology Fact", "Physics fact", etc. In fact, when one looks at the history of science, many facts and discoveries could have readily arisen in other fields than the ones in which they were discovered.

The facts have no labels that purely aim them at one discipline or another. Many breakthrough discoveries were made either in one field and found its widest application in another or as the efforts of cross-disciplinary science.

Modern liquid chromatography can be traced to the advent of bonded non-polar phases. Prior to that, much of liquid chromatography was done with polar adsorbents. Non-polar analytes were limited to separations on non-polar oil or wax coated columns. These were not well unsuited for aqueous separation media. The huge number of application in biological and pharmaceutical chemistry that now are done could never have arisen without scientists who used silanization reactions. These were not developed for coating silica, but rather were developed to give glassware that did not adsorb trace levels of polar compounds. Someone recognized the potential in the other area.

An additional incentive besides innovative discoveries is the current economics of research. Funding is much tighter today in all three venues. If scientists approach these cross-disciplinary problems from different directions, but in collaboration, greater results can happen. It is easier to fund two smaller research projects done in concert than it is to fund one as big as both. This is true both financially and technically. Those who approve budgets do not necessarily accept the coordinated efforts of two disparate research projects. This novel innovation is harder to accept and seems riskier. This less-likely-to-succeed image makes funding approval lower. If the researchers, however, sell the two ideas as separate projects each aiming at only a part of the overall goal, then approval is more likely. This seems illogical and convoluted, but today's funding mentality is much more constrained than in the past.

2.9 The Tools (Part 1) – Tools and Mechanics of Research: Putting Together Your Toolbox

63

This is a rather diverse chapter. The various topics are too small to warrant separate chapters, but too important to overlook. I decided to gather all these sorts of areas into two sections. Their diversity and number grew. Each in itself might not be key to success, but when added to the others that are described in individual chapters, you will increase your chances of success. Leaving these out diminished the chances.

There are several habits and attitudes that are involved in the actual doing of research, the work in the laboratory or on the computer or in whatever venue your research may take place in. These are not the more wide-ranging ones that make an individual want to do good research work. Those, like curiosity or skepticism or the drive to learn, are discussed in separate sections of this book. The traits in this section are ones more basic to the tasks needed to do the work. These help turn the thoughts and attitudes into successful laboratory work. There also are some things that are very wide-ranging that they are reiterated here as general characteristics in order to reemphasize their importance.

The skills, ways of doing things, and tasks are not

absolutely necessary to doing good research, but they can make the research efforts better. In looking at the ways many scientists do their work, these things kept being brought up in the conversations. They make good research more likely and easier. Many researchers cited examples in their work where they did not do some of these, had difficulties, and learned that these things are important. They do not guarantee success and are less important than the critical, innovative, and wide-ranging thinking or the personal skills. So mastering them or not is still a choice, but successful researchers generally choose to sustain their successes.

Initiative

There is some truth to the saying "If you want it done right, do it yourself." From the standpoint of your own research interests, you can go into new areas and apply new things that improve your own work.

In general, initiative is a key. If something must be done that you can do, the first option can always be to do it yourself. If entrusting a task to another is better for a teamwork mentality or because you cannot do it, then you still must take the initiative to make sure it gets done. Passing it on and assuming it will be done does not ensure success. Some people need reminders or schedules to prompt them in a task. Give them those when it is needed. If there might be friction in reminding, do it in a more indirect fashion, such as discussing the next steps that rely on the task being done. You should remember that something important to you is not necessarily as high a priority to anyone else.

Be prepared to send out follow-up phone calls and e-mails or written correspondence whenever you have had a discussion with someone about plans, commitments, and schedules. Reiterate what you had discussed to ensure both, agreement on what was discussed and to create a reminder that tasks need to be done. For reminders that you cannot deliver in person, telephone calls are much better than a letter or an e-mail. Those can easily be ignored, even if only because the person is busy.

You will find many valuable people to form your network who will help you in many ways. You must remember that they cannot unless you ask for specific help. They will also often need reminding. They all have busy schedules, too. Their schedules and priorities are theirs, not yours. So your favor may not be at the top of their list of things to do. It will certainly be lower than it might be on your own list of priorities. You are the one with the most to win by taking action and the most to lose by not. Ask your network and manage it enough to get it done (without being too pushy and annoying – as it is a favor).

Meticulousness and Fastidiousness

The laboratory work for an undergraduate degree emphasizes precision, accuracy, and reproducibility. There is a major reason. These qualities must be learned until they are an inherent part of doing laboratory work. Unfortunately the emphasis is not always sustained once the work is not being graded for these qualities or when the gaining of a degree is not the goal. Care in doing laboratory work makes the assessment of results and the variables that affect them much easier. If the making up of stock solutions or the setting of a thermostat are not carefully done, then how can the influences of concentration and temperature be assessed properly?

One famous example of erroneous reported research comes to mind. Polywater was hailed as a new, never-before-seen variant of water that behaved in many ways like a very high-molecular-weight substance. It was ascribed to a polymeric form. Later work could not replicate the initial reports. There was, however, evidence that trace-level impurities could account for the observed behaviors. Insufficient cleaning of glassware was supposed to be the cause. There are innumerable examples of less visible errors that result in later retractions or corrections.

These traits are especially valuable when the laboratory is used by several people. Orderliness in a very busy laboratory can be the difference between the researchers getting along or not. Common resources or what might be called communal areas are a contentious point when some are not fastidious. A messy person, especially one who leaves a commonly used piece of equipment in disorder, will soon rankle the others in the laboratory. Few things build up such tension as someone who leaves a mess around a balance, on extraction glassware, or in a centrifuge, oblivious to the inconvenience of others or even worse, expecting the others to tidy those messes. The one thing that might be more irritating is when valuable time is spent searching for chemicals, tools, or an apparatus that are not returned to their proper places. Repetitions only add more and more fuels to the fires of resentment the meticulous and fastidious ones have for their messy counterparts.

One solution which I found useful as a laboratory supervisor is one I learned from my graduate school research advisor. He required a monthly laboratory cleanup that everyone had to participate in without exception. Rearranging and reorganizing things were a shared task. He also had the prerogative of calling for a cleanup whenever he thought he saw disarray or clutter. This also kept everyone aware of being neat and returning things to their proper place as no one wanted to be the genesis of one of these several-hour-long cleanups. The psychology of group cleanup efforts eases a lot of the tension felt by the meticulous ones for the messier others. They see those people cleaning up and putting straight areas that would have been a contentious issue.

One caveat must be given as far as thoroughness in experiments. The inclination of some might be to do experiments until the results are absolutely confirmed. This may be a good principle, but can be overdone. Assessing every variable in minute increments or performing a large number of replicates carries this too far. There is a point where the gains in precision, accuracy, and the other measures of reproducibility are not worth the further efforts.

Planning, Organizing, Scheduling, Budgeting, and Running Meetings
There are many steps in a research project. Some must be done simultaneously, others require equipment or other resources must be scheduled and chemicals must be acquired or made. Planning these allows a smoother flow and coordination of all the steps. The alternative is to stop and wait for the arranging of each one of these as it is taken care of at the time each is needed. Timelines and flow charts are good tools.

These tools help an individual chart out what tasks need to be done when. They also help to prioritize when work on different projects must be done in the same time span. If they are detailed down to being daily "to do" lists or schedules, then they make multitasking easier. These often are done on a daily or weekly basis with changes made as things arise. For some, there is a sense of accomplishment as an added bonus when they can mark off a completed item and an even bigger one when they can (infrequently) end the day with all of the planned items done.

Several people highlighted that they carry a small notepad and pen with them in order to note any thoughts of a thing to do. These ideas can pop into the mind at any time, even in the middle of the night. I, among many other people, keep those two things on the nightstand for quick noting at the moment when it is fresh so that a brainstorm or a needed reminder is not lost. Even the most brilliant and lucid ideas can be forgotten by morning. Writing down the ideas and whatever important details have come into your mind ensures that they are not lost. There have been times in writing this book when I thought of very good points to make in certain exact words that will not last as only thoughts. The sheet is then taken and added to the main list of things to do that is in the office or laboratory.

These up-front planning steps become very important if many individuals are involved in a project. Some people may have schedules that only offer windows of opportunity for them to do the needed tasks. If subsequent steps cannot be done without one of these individual ones, then there is a bottleneck. A lack of coordination may result in a very long delay for everyone's work. Sharing these timelines, plans, and schedules also makes sure that everyone knows their roles, the timing, and what needs to be coordinated with others.

The comment from someone who was less than a year into a government position after receiving her doctorate was "the real issue I had was all of the bureaucratic levels that need to be dealt with in order to get anything done". Her experiences were echoed by others in industrial positions. Issues of "performance measures, milestones, and strategic planning" now are commonplace in both of those venues. Academicians writing proposals must describe the applicability and values of their research to society. Financial planning and budgets are another bane of today's research scientist. Keeping track of expenses is part of doing research. The individual researcher is now in the same type of situation as the entrepreneur or venture capitalist. Research is not freely funded in any venue; it has to be sold to funding sources. These sorts of issues are not learned in graduate school unless the research advisor exposes the student to them.

Another very important aspect of planning, organizing, scheduling, and budgeting is meetings. At some point in your career you will need to convene a meeting. The first step in planning for a successful meeting is to think of the topics you want to have covered. Write these down as a draft listing. Think of the order of the topics. Are some better placed before others? Include both an introductory and a conclusion/ summation section. The first sets the tone and agenda you wish and the final covers what was discussed and what was agreed upon. This ensures that no one leaves unaware of responsibilities or with different impressions of what will be done.

Next, estimate the time you want for each topic. Include some extra time if discussions are to be held for each topic. If your total time is long, you must decide if this is manageable. If the meeting is too long, people will not be able to fit it into their schedules or it can be too draining to be effective. If you cannot pare off topics to shorten the meeting, then split it into two. Create two separate meetings that each deal with a specific set of the topics.

When the agenda is done, you can put together a list of attendees that you want. Subdivide this into critical and optional people. Critical ones are the people who you think will either be involved in tasks agreed upon or who can make the decisions that must be made. If scheduling is a potential problem, contact each critical person to coordinate a time for the meeting.

The meeting mechanics are critical for several reasons. Keeping focused on the topics, staying somewhat on schedule, and preventing clashes when plans or opinions are discussed are some of the keys. I will describe some of the practices used in the industrial venue since it seems to deal with meetings better than the other two. In conducting meetings at conferences, several academicians have thanked and complimented me on well-run meetings which seem to be a rarity in that venue. Industry emphasizes effective meetings to the point of often offering classes on the subject.

Well before the meeting, send out copies of your planned agenda so that everyone knows the topics and is prepared to discuss them. This allows for feedback on adding topics or attendees to the lists. This also allows everyone to add the meeting to their calendars.

When the meeting is held, you, as the convener, should set the tone by introducing the agenda. If it is possible, a large copy of it can be made for everyone to see during the meeting. This allows you to refer to it whenever you need to and reminds the participants of the topics at hand and the schedule. First ask for everyone's agreement to the agenda or if there needs to be additions or changes. Once the meeting begins, you should act as the moderator, reminding people of the topic being discussed if they wander from it, keeping the discussion moving, drawing in all the attendees as participants by asking questions or for opinions, and acting as a mediator if disagreements get too acute. You do not want an open-ended round-table discussion, so bring the discussion back to the topic being discussed if the discussion becomes tangential.

An additional role is that of recorder or scribe. This is a person, either yourself or another from the optional attendees if you think you will be too busy and involved in the meeting, who writes down the progress of the meeting. This can be done with pen and paper or on a laptop computer, but is best done on a large notepad (a flip chart) that is readable by everyone. This allows you to review the meeting better and to gather the ideas. In addition to the description of the topical discussions, the scribe can keep lists of issues that arise that need to be addressed but are diverting at the moment and tasks to be done.

The meeting has purposes that you planned for, so it is your responsibility to keep everyone's attention on those. When every topic has been discussed and all the agreements that are possible have been made, it is time to conclude. Summarize the meeting and highlight the main points. These include things agreed upon, plans for actions and who will do each, things to check into and who will do each, and the initial planning for a follow-up meeting if one is needed. Much of these can be reviewed simply by going over the scribe's notes. Remember to thank everyone for attending. Afterwards, copy the scribe's notes and distribute to everyone.

The Personal Library and Databases

The personal library does not refer to the shelves of books in your office that contain the key books in your research area. These are very useful and cannot be overlooked. The personal library is the polyglot of articles and bits of information that are much more widely scattered and therefore more difficult to assemble and organize into a useful form.

70

There is a huge amount of published research. Keeping up with it is a chore. A second chore is storing it for future reference in a fashion in which retrieval is quick and easy. Each researcher must assemble a personal library to support the specific research interests that are worked in. No one book, or even a small number of books, has all of the relevant basic and background information. If the researcher remains diverse, then this is even more so. How do you create your own "Dewey Decimal System" (That arcane notation of letters and numbers on the outer spines of books in a library)? One that works for you and the ways you think and do your work.

Folders and file cabinets are nice, but after the saving of a few hundred papers they will become a muddle. This is unless there is a good system of storage that you can remember the rules of. People use different ways of thinking into categories, so the organization must suit your own ways of thinking.

This may seem to be a trivial topic, but after only a few years of research, a person's personal library has many references. The mind may remember a few details of the work and who did it and where it was published. This is neither enough to base one's work on nor enough to cite in a manuscript. The spending of a cumulative time of hours and hours of sorting through files is a very inefficient waste. The time is better spent on doing the research or writing.

Certain papers may fall into several useful categories. For example, a paper may describe an HPLC separation of the extracted bark of oak leaves of different species in which mass spectrometry, NMR, and IR are used to identify components. If the extraction, HPLC, and various spectral techniques each involve something novel, then where would this one paper be filed? Many researchers create folders specifically for such multiple topic papers and note on each the featured areas. Others make multiple copies, one for each specific topical folder. Others will have a cross-referencing index sheet that tells them that a paper in such-and-such folder has information that also fits into the other where the note is.

Occasionally reviewing the personal library's arrangement is necessary. New categories may arise or old ones get so full of references that subdivision makes sense. The system should have categories that quickly and efficiently help you find the references you want. When you pull out a file folder or look into an electronic equivalent and notice how hefty it is, make a note to review it for possible reclassifying and subdividing. In my own case, my original folder on "spectroscopic data" expanded with time into folders on UV, fluorescence, proton NMR, carbon-13 NMR, infrared, and a miscellaneous "other spectroscopies" folder. It is much easier to find a specific article among lots of thin folders that are organized well than it is to find it among very thick folders. It is also easier to organize lots of thin folders than it is to internally organize a thick one.

The inclusion of references to information in books is a dimension of information that does not fit into this filing system. Although some book sections and chapters can be copied and included in the files with journal references, there are still going to be some very useful information in the books on the shelf. Certain books contain so much useful information that they must be at hand. Is there an easy way to make information in books fit into the filing system?

I heard of two approaches that would. One is to include a sheet in the file folders that becomes a listing of pertinent parts of books. Each listing has not only the book and useful page numbers, but also a synopsis of the information there. The other way is to create such listings and keep them within the book itself as an additional cross-referencing method. The result of using these would be listings to all relevant book pages in all the appropriate file folders. A book, therefore, might result in listing on the synthesis of compounds or their spectra or their analysis by chromatography or their health effects in each of the folders containing the analogous journal information.

Some types of data can be spread throughout hundreds of papers and books. Creating a personal database makes these convenient. For example, certain information is routinely contained in reports of newly synthesized compounds.

Melting point, crystal structures, colors, and UV, IR, NMR, and other spectra are some of these. Sorting through these many sources to find specific information is time-consuming. Copying and collecting these in a notebook or another analogous way is useful.

I collected a UV spectral library for the polycyclic aromatic hydrocarbons (PAHs). It consisted of an index card for each of the PAHs. This was done by first either photocopying literature spectra or those that I had collected in my own research. After shrinking down to the right size to fit onto a certain part of the cards, each went on its own index card that also contained the empirical formula, the compound's IUPAC name and any trivial, alternate ones, and a reference to the literature spectrum's source. Thus, a small index-card file box contained a unique collection of hundreds of spectra that no other researcher in the field had at their fingertips. It made spectral matching easier when I later did work on HPLC with a full-spectrum UV detector for analyzing PAHs in a variety of complex mixtures. Flipping through these note cards was quick and easy.

These handy sources of data, if maintained diligently, are a personal resource that gives you an edge over anyone else trying to do similar work. It also makes preparing any manuscript, but particularly those for review articles and book chapters, much simpler.

Record Keeping, Notebooks, etc.

Much of research is done in the laboratory. This is a notation of the actual combining of reagents, the synthesis of intermediates and their isolation for subsequent steps in a reaction series, the preparation of samples, and so on. While much of today's instrumentation can record work electronically as it is done, there still is a great part of laboratory science that requires the old-fashioned taking of notes. This might be done on a computer instead of by pen or pencil and paper, but it still must be done.

This is best done often and to the point of being habitual. Writing odds and ends notes on scraps of paper is one way, but it is not systematic and organized. Some of those little slips of paper may be lost or misplaced. Keeping a laboratory notebook is one solution that is organized and if used correctly it is also centralized, containing all of the important laboratory notes in one place.

If you resort to using paper, remember that readability is a factor. Take the time to write clearly and legibly. I must confess that this has been one of my own personal weaknesses, but it is embarrassing when reviewing an experiment and I am unable to read my own scrawled writing. Be careful with your notebook, as a spill of water can turn those precious notes into indecipherable smudges. Include all relevant calculation, written out in full. This allows you to check your equations and mathematics. If you use a literature value, note it down with its source information.

It is also reasonable to make a photocopy of laboratory notebook pages that contain uniquely valuable data or to have some other backup system in place. As with electronic files, copying and backing up often and storing the extra copies in a different place than the original protects your work against loss or accidents.

Carry your notebook wherever you are doing work so that you write down all of the details, every weighing, every dilution, every reagent and its pertinent information, etc. Some scientists include chromatographic and spectrometric printouts in their notebooks, folding them so they fit and attaching them in a way that makes them a part of the notebook. If you use special apparatus or glassware, make an illustration or photocopy the schematic or figure in a publication that was its basis. Tables of reference data can be handled similarly. Always note on these figures and copies what they are and what their full source information is. Anything of this sort that is useful should be included so that you do not need to sift through numerous books and articles when you are ready to write a report or manuscript.

Having all of the laboratory work located in one place makes finding any information much easier when a report or manuscript is written. One synthetic organic chemist said that he set aside a few pages in the back of his notebook. This became a listing of what was on each page, an index that helped him quickly find specific information.

One other aspect of record keeping and notebooks is that these can be absolutely precious when there are conflicting claims for discovery. In the past, this mostly meant the accolades

73

for first doing a certain technique or finding of a new compound or fact. Today the legal aspects, particularly in the granting of patents, may really be worth a king's ransom. The inclusion that an idea was thought of on a particular date, with later entries for the dates of subsequent work, can prove this priority. This is the main reason industrial notebooks are generally notarized and involve very formal procedures.

Data Sharing, Archiving and Retrieving

For many projects, data that is collected, is statistically analyzed, and is interpreted must be shared among all participants. This may involve using common computer protocols and software. This may seem unnecessary, but if more people participate or understand these steps, then there will be the possibility of more interpretations and less of deciding on an erroneous conclusion. If the research is compartmentalized, with only each individual being involved in and responsible for a certain part, then one person's opinions can misdirect the whole effort.

The collection and collation of data can be difficult as the programs used to generate the data and the formats of presentation are often unique to each instrument. Even instruments using the same operating programs may use different data formats and protocols. Easy statistical treatment becomes difficult in any uniform fashion because the various pieces of data cannot just be electronically transferred. Each participant must learn the various formats and some rudiments of interpretation. This breaks down the compartmentalization. If everyone has some grasp of the process they can more readily notice errors due to formatting, unit conversions, decimal errors, and other causes. They do not rely solely on the data person, the statistician, or computer programmer to look for such errors. For example, those people might not understand that absolute temperature is used in one type of data and not convert the measured C temperature.

After the experiments are done and written up in a report or publication, one last step must be done. The pertinent data and notes must be stored. A good scientist creates a system in which these are archived and easily found in any future need. This does happen. When experiments are continued after a pause or when other people must perform the continued work (a common situation in academia when graduate students finish their degrees or in industry and government labs when someone transfers or retires).

Any important reference papers should be copied so that they are kept in association with the research work that relies on them. All of these copies should contain the full source information in the form as it is used in publication citations. Some journals and many book pages do not contain this information, so it must be added to the copies so that subsequent workers know what these are and where they came from.

For some types of work in industry and government laboratories, these types of archival record keeping are mandated by regulations.

Commitment, Integrity, and Trustworthiness

These characteristics refer to how others perceive you in terms of doing what is expected of you when you are being relied upon. In terms of research career success this leads to either more or fewer opportunities. Being good at it opens doors, while being bad will keep some shut. You might think of it as another facet of ethical behaviors. It is emphasized here because doing it may be the right thing, but not doing harms your reputation and that greatly lessens your chances for certain kinds of success.

Being reliable is a key to success in many of the areas of research that are discussed in other chapters. Any collaboration, team effort, or other project that involves more than one person, will make this necessary. When others rely on the quality and timing of your efforts, you must set a priority to meeting what is expected of you. This is doubly true when you have agreed to the criteria and schedules that are needed. When you do not meet these agreed-upon things, you lose credibility in the future. If your unreliability creates any problems, then you may not be chosen for those later team efforts.

You must remind yourself that what might be your own priorities in work scheduling must be redone and rethought to take into account their importance in the work of others and to the overall success of a team. Sometimes an individual may get focused on what is more important to him or her. This loses sight that several others are waiting for that work to be done or for an instrument to become available.

Being involved in some of the formal parts of the scientific community – the societies, the committees, the journals, and the conferences – brings with it many duties. For most roles there are definitions of the obligations. For certain roles there are responsibilities that are not explicitly defined, but which are implied from precedents. I think of certain situations that have arisen in my research career. Someone in a role like being the chairperson in a symposium session or the guest editor of a journal issue is expected to assemble the roster of participants and remind them to be prepared.

An officer in the society is not only to do the nominal things like recording meeting notes if you are the secretary or collecting membership dues if you are the treasurer, but also be an ambassador for the society. You make it more visible and try to bring in more members. There are sadly very many people who are willing to accept these roles for the prestige, but who do not want to do very much to fulfill the expectations. Your reputation can be hurt far more than any prestige gained if you do not earn it. The scientific community may seem large, but just as in society as a whole, word spreads rapidly with both good and bad repute. The bad especially has more effect because people remember it more strongly.

In contrast, if you are perceived as being reliable others will offer you collaboration opportunities or mention your name as a nominee for work for journals and societies. Your reputation extends your network as your immediate contacts speak well of you to their contacts. If you are thought of as trustworthy, then other researchers will seek out your opinions on their developing research or on results where the interpretation is unclear. These types of interactions can be enjoyable and stimulate other research.

2.10   The Tools (Part 2) – Handy Tools, But Not Always Needed

The variety of topics discussed in this chapter are ones that I think are helpful in doing good research, but are not as essential as those described in the previous chapter. Using the toolbox analogy, these might be those lesser used tools that are not used on every project. They, however, are useful or even ideal for certain ones. The earlier chapter dealt with the main components in the toolbox, analogs to the screwdriver, saw, or tape measure. This one deals with the calipers, the level, and the adjustable wrench. There are several areas that are necessary if you do research in a certain topical area. For example, working in many fields of analytical chemistry can be easier if you have some mechanical skills or understand statistics.

Statistics

What is your reaction to the following statement? "One out of every twenty of your data points will be bad." This pronouncement sounds harsh, but that is the truth if you use statistics with a 95% confidence level. On the average, a 95% confidence level means that 5% will fall outside of those limits. This was pointed out to me in a class on statistical methods. I later learned its nuances when I taught a class on experimental statistics at a nearby college. Statistics does not allow for you to throw those points out. How do you really define an outlying piece of data? There are specific tests to see if a data point is truly an outlier and can be discarded. Too often researchers either use statistical tools incorrectly or not at all. The resulting data may then be flawed.

These are not trivial questions if truly top-level research in certain fields is the aim. Analytical and environmental chemistry rely heavily on statistics. Poor understanding or use of statistical methods can be a "red flag" for the knowledgeable reviewer of the data. It may indicate sloppy or unknowledgeable experimenters. This may occur in the review process of a regulatory agency assessing a report or in a manuscript that has been submitted for publication. In either case, the researcher's work will not be looked on in as favorable a light as if the statistics were done properly.

Computer Programming Skills

There are fundamental computer skills that every researcher should have, such as being able to write using a word processing program, being able to assemble data in a spreadsheet or database, and creating simple graphs and tables. In some fields such as computer modeling or the complex statistical treatments of chemometrics, a deeper understanding of the programs is needed. These specific usage programs may not be key to everyone's work or to every project, but being versed at least in what they can do helps in planning a project.

In compound analysis, the identification of components in a mixture using chromatography with certain hyphenated spectral detection techniques is accomplished by spectral matching. If you understand the algorithms that are used for the comparison between unknown peaks and standard spectra in a library, you can then be more astute in recognizing errors. For example, UV spectral matching is most often done on a point-to-point basis. This works well for unsubstituted compounds, but there will be a red shift in the spectrum of 1 to 2 nanometers for each addition of a methyl group. The program may not recognize this, but looking at the spectra will readily show the characteristic UV band patterns of the parent compound. You may be able to devise a simple change in the program to shift the unknown's pattern slightly downward for matching.

This is an increasing need at this time as more and more work is based on computer models and simulation. Using calculated values instead of experimentally determined ones is now common. As an example in the environmental area, toxicologists routinely use computer-generated values of aqueous solubility and the octanol-water partition coefficient to assess the effects of chemicals. These rely on modeling the compound in question and using programs to calculate those two parameters. Few of the people using those programs know either the base set of compounds used for the model and how their target compounds might differ or what assumptions go into the models. When a model relies only on certain types of chemicals, then ones of much differing structures may not behave under that model.

Understanding the models helps produce valid results. If, in addition, you have the programming skills to modify the model to better suit your use or the compounds you are studying, then you can do valid work in areas that another researcher might not be able to do. You can also modify the programs as you find new things, improving the models and algorithms through improving iterations.

Mechanical Abilities

"Chromatographers are scientific plumbers" I was told in graduate school. The use of wrenches to change columns or to couple injectors and detectors can be simple, but in order to do it properly may be more involving. Minimizing connecting tubing, putting connections together by choosing the right fittings, and finding leaks or blockages are learned skills.

Anyone dealing with pumps learns how to repair certain things. For an HPLC it may be check valves. For a mass spectrometer or other low-pressure system it may be the vacuum pump bearings or a belt or a sample valve gasket. Spectroscopists learn how to align optics or clean windows where sample or contamination can collect. Almost everyone learns how to check and change fuses because every instrument has them.

Unfortunately this is one skill area that seems to be diminishing. Several professors pointed out that it is much less in graduate students now than in those of a decade or two earlier. Some of these researchers supposed that this was because the things in our society no longer can be tinkered with to see how they work and for repair.

Appliances, clocks, and cars were once learning places that could be disassembled, examined, and reassembled. Now they are digitally controlled and many parts are either too sophisticated or inaccessible for this. Repairing an automobile once only required a simple set of tools. Now it takes a specific computer-controlled monitoring system or an oscilloscope to do the same things. This may not seem to be a critical issue in doing research, but one professor pointed out that beginning academicians must assemble much of their equipment because they cannot afford to buy everything off the shelf and ready to work. A young professor in spectroscopy or another instrument-oriented field must have mechanical skills and have students who either do, too, or who can be taught it without taking up too much of the professor's time.

How does a young scientist gain these skills if there are fewer opportunities? There are still some chances of learning by doing. Look around for them. Become familiar with the instrument manuals for key pieces of equipment in your laboratories. Take the time to watch someone who is repairing a piece of apparatus and ask questions to ensure your understanding. If a serviceperson is repairing something or even doing routine maintenance, watch and learn. Even if that piece of equipment is not important to your own work, learn the techniques and approach to opening up the instrument, checking out its parts, and the cleaning of it, the replacing of parts, the reassembly, and the final checking of performance.

This is especially possible in a graduate school research group where the more senior students and postdocs do know more of these things and the tricks learned by experience. Be a sponge and absorb their knowledge when there is any opportunity. Be aware of times when equipment is being serviced or repaired. Watch and learn. Ask questions. Take notes.

Language Skills and Cultural Awareness
The sciences, particularly physics and chemistry, were once multilingual. Much work was published in German and significant amounts were also published in French and Russian. English has grown in the past two decades to be the common language of most scientists, being either the primary or secondary one for almost every researcher. Although the others

are less critically needed, they are still useful. This is especially true in the disciplines of organic and inorganic chemistry where key references such as Beilstein's Compendium and the Gmelin series are in German.

Additionally, being multilingual is an asset in networking with those in which English is a second language. I only remember a few bits of the German, Spanish, and Japanese that I once learned in classes. Even with this limited grasp of these languages, my being able to offer the pleasantries and greetings has shown me the value of speaking more than English. The surprise and smiles when I greet someone in their own language creates a closer contact.

Another aspect of this sort which I have personally seen is that knowledge of other nations, and their cultures and histories, also makes a connection. This knowledge also is very valuable on those times I have traveled to other nations. These personal connections lead to friendlier conversations and more open exchanges of ideas. They often lead to true friendships in addition to working together.

In order to gain some cultural awareness, it is useful to read about the history, society, and current happenings in other countries. Even bits and pieces of this type of knowledge can make a good impression when you interact with people from other nations. I have found that people from many nations, particularly those from outside of Europe and North America, are pleasantly surprised if you know about their country. I learned this because I have an interest in geography and world history, but the results are the same for anyone who takes the time to become familiar with other places.

This should be a more important effort if you are actually traveling to another nation. Reading some background information will make you both more aware of things you encounter and in some cases of things to avoid. There are numerous cultural taboos and impolitenesses that can be done by the unknowing visitor, but conversely knowing of them and not doing them avoids an embarrassment or may even make a good impression. As an example, the giving of business cards is a very important procedure in certain cultures. The business card states a person's position and is a status symbol. Receiving it in that light is considered good manners, while not may be considered as being rude. Americans do not hold this view and can make a misstep by casually accepting a business card and quickly putting it in a pocket without looking at it carefully as is the custom. Some cultures even have almost ritualistic stances in offering and accepting business cards. In Japan, they are offered with both hands during a slight bow.

Understanding cultures also is applicable to working together. This is also touched on in the chapters on diversity and team work. The culture in the United States and Australia, among others, is highly individualistic. Other cultures are not. Japanese culture, for example, is very cohesive and stresses the sameness of people. Some cultures are very stratified. Scientists with advanced degrees are looked at as being on a different level than those who do not. Formalities are always observed with the use of the title Doctor or Professor.

Mining, Digging for the Rare Nuggets – Being Persistent in Your Work
In several examples from the successful research projects that others described and in my own experience, one of the things that made the research more innovative or valuable was relying on something that others in the field had not been able to use. This might have been finding unique model or standard compounds, an ideal sample, or an obscure reference, or just taking one known fact and extending its use into new areas.

This might be called being persistent, being thorough, and being diligent. The researcher, who does the combination of those three things, will be in the situations where research can be done that no one else can do. Doing the extra things will open doors of opportunity. This is created by a mindset that does not stop at the points of preparation where others might start their work. The average scientist sets a target as the point in which work can commence. The good researcher does this and then prepares more. This does not need to slow down the pace of the research, as many of these things can be done continuously as part of the overall task of keeping current.

In each of the cases that were described by the successful researchers, they all scoured the common sources and came up with whatever was available. This did not seem to be enough and so they kept looking in the lesser known places. This added effort, by not stopping when the initial search seemed to show completion, opened upon opportunity. Lesser known journals or those in other fields became sources of information. Small, specialty chemical companies or a professor who had made a compound for unrelated research are sources for some very odd, but potentially valuable, compounds.

There are small chemical companies that focus on supplying one or a few compound types such as low-molecular-weight polypeptides, alkenes, carotenoids, multi-ring thiophenes, porphyrins, or polycyclic aromatic hydrocarbons (PAHs). Through in-house synthesis or working with those synthetic organic chemistry professors who make some of these compounds, they fill a niche market. These sources are lesser known and must be sought after since they do little advertising. Looking at the sources in publications is one way to find these small businesses. Another is networking. Many researchers who might know of such a company for compounds of interest in their field assume that since they know of the company then everyone else must know of it. So they do not spread the word and only acknowledge this rare source when they publish the work. Ask your colleagues if they know of any of these small-company sources.

For some researchers this searching meant keeping one's eyes open and being aware of the potential for certain compounds. When one popped up, the researcher obtained it. Even the large and well-known chemical suppliers have a few odds-and-ends compounds. Looking through the catalogs can make you find very valuable compounds. There is one very large supplier of fine chemicals, Sigma-Aldrich, which has a collection of thousands of these sorts of odd compounds. Perusing that catalog, which is conveniently available as a free compact disk, shows many compounds that are not available anywhere else.

In the catalog from a major supplier, I found three polycyclic ketones that were not available from anyone else. These could be reduced to my target eight- and nine-ring PAHs in one reaction step. My scouring the sources gave me three useful compounds that others in the field did not have. This resulted in one paper and a large supply of these three PAHs that I traded other researchers with to get approximately a dozen other PAHs. This bartering among researchers is also another way to scour nuggets. A professor or researcher may not want to sell his or her valuable compounds, but will readily trade them for ones that are useful.

In a similar fashion, others scoured the literature and looked at papers in journals in seemingly unrelated fields to their own interests. They looked to see if any aspects of the work were applicable. For example, a chromatographer might read the literature on zeolitic catalysts in the hopes of finding references to permeation behavior for certain molecules. In doing so, other potentially useful facts might be noticed.

These compounds, facts, or samples may be available to anyone who looks in a certain fashion. Very few researchers do this sort of searching in addition to what is the norm. Those who do, however, create ready opportunities for innovative work. This off-the-beaten-path searching gives them access to research that others cannot do without those hard-to-find nuggets.

How do you set yourself up to do those? Think of everything as potentially relevant to your research interests. As the cliché goes "Think outside of the box." Do not gloss over anything and think "This is of no interest because it is too far from my own work." Look in papers for details that do connect. A very important aspect of this is that these searches are extra. They can be done piecemeal, here and there in little bits that fit your schedule. If you do not have time now, they get deferred without impacting your core research plans. If you look in those bits as you can, though, you then increase your opportunities for doing things others have not done.

Is there an odd solvent used to dissolve a sample that may be a chromatographic mobile phase, even if it is exotic. This is how I used chlorobenzene and toluene as mobile-phase components in the first separation of the higher fullerenes. No one had ever used those solvents with a reversed-phase column, but those solvents were very good at dissolving the crude fullerene mixtures. It was my thinking "Why not?" rather than thinking that reversed-phase separations had mobile phases that were more polar than the stationary phase. Taking the fact and using it differently was the key.

In other research, I relooked at results for some syntheses and wondered why only certain isomers were reported. The accepted interpretation of the published synthetic work was that the reaction mechanisms were selective so that some isomers were formed and others were not. Performing the same syntheses and using better separations, several "new" isomers were found in each of the various reactions I looked at. These compounds had been made in the earlier work, but missed and only the "easy-to-separate" isomers had been isolated and reported. What was innovative? It was not my use of modern separations techniques, for many other researchers might have done that. It was my thinking that the thirty-year-old papers might have missed something and not thinking that the mechanisms of the reactions were selective. Knowing the limitations of the earlier separations made me aware of the potential of the better, more modern ones.

3   Non-technical Areas

3   Non-technical Areas

3.1   Communicating

3.1.1   The General Common Points

One of the most important parts of being a successful research scientist is to have good communication skills. Although some people have natural talents in speaking and writing, these still must be developed in order to be fully effective. A third skill is listening, which is often overlooked. These chapters deal with some of the skills and mechanics of communicating. This introductory one is both an overview and a guide in how to use communications effectively.

For everyone who must communicate in writing and speaking, two very important tools are the dictionary and thesaurus. In order to use them better, you should get in the habit of using them often. Whenever you encounter an unknown or unclear word in your reading or listening, make a point of noting it. Then look it up as soon as it is convenient. It is best to do this as soon as possible after seeing or hearing, as the connection between meanings and contexts are freshly made for each unfamiliar word.

When you struggle for a word, remember to use the thesaurus. It was created to give you options for more meaning or less redundancy in word use. Science naturally does this by using less familiar words often. We synthesize new compounds when we make new ones. We collaborate when we work together. Agglomerate, conjugate, stratify, scintillate, disassociate, and permeate are all common scientific verbs. These are among the dozens and dozens of lesser known ones to the general population that we use. Expanding of the vocabulary is only natural in science communication and is done with every new concept and discovery.

Make an effort to learn more words. This is especially true when you hear words that turn out to exactly express an idea. Note them and assimilate them into your own use. The differences in the meanings of words can be important in clearly passing on an idea. Scientists deal in precision and efficiency so exactly passing on your meaning in fewer words should not be an alien attitude.

Another aspect of vocabulary is to use it completely. Linguists point out that most people know and recognize many more words than are in common usage. These less commonly used words often are more specific in meaning or more vivid in imagery. Do not limit a message because the chosen words are only the ones that are commonly used.

As a writer you should remember that redundancy in the text gets boring for the reader. Using the same word over and over gets trite or mundane. Varying your words does not need to mean choosing rare and unknown ones. People know and read many words that are not commonly used. In fact, the mechanics of reading allows for this. A reader in seeing a word thinks of its meaning. The reading reminds her or him of the meaning much more readily than if that word had to be thought of for speaking or writing.

Do not avoid metaphor, analogy, or other figures of speech if they create the mental images you need in order to transmit your ideas. One of the following sections deals with writing styles and why many scientists narrow their writing and reading so that personality and good reading are often left out. This leaves boring and dry descriptions.

The More Difficult Tasks – Communicating Well in a Second Language

The ideas in this and the next chapters on speaking, writing, and listening are for the situation where the communication is in the primary language. This, however, is often not the case. With the rise of English as the preferred and accepted language in most books and journals for written science and at most conferences for spoken science, many scientists must gain the skills to be good in these situations.

What are some things that can be done to aid in better communicating in a second language? The saying "A picture is worth a thousand words" comes to mind. Although the phrase may not be accurate, it basically points out a strength. Relying on the figures and tables for a publication or a presentation can be even better in this situation. Good illustrations in a manuscript require less text for explanation. This is even more true for a presentation. A non-native speaker who is unsure of the language may present more slides (or other visuals). Each slide would contain more written information that a native speaker might convey orally. Thus, a non-native speaker may present research with more slides and only have to describe each briefly as compared to the same presentation given by a fluent speaker. These visuals can be prepared carefully and edited or redone numerous times until clarity is reached. A spoken message cannot be as finely tuned or offer fewer chances of mistaken delivery as the same information given on visuals.

When you are preparing a manuscript or presentation in a second language, you should find fluent or native speakers who can review the work for its correct language. This is easier done with a manuscript than a presentation, so the content of a manuscript should never be submitted without a review for language correctness. If there is no one in the research group who is comfortably proficient, a copy of the manuscript can be sent to someone in your network who is proficient. Correctness in language – the grammar, syntax, and meaning – can be reviewed and corrections suggested. This makes the manuscript much more presentable to the journal's editors and reviewers. For a presentation, practice that is strictly for the correctness of the language can be done with a language-proficient colleague. As described in the chapter on speaking, do not script the presentation. Use notes to remind yourself of correct language to use in describing certain things. Brevity is not bad if you have good visuals to help deliver the message.

Using Your Communications Skills

After you master writing, speaking, and listening, what do you do with them to bolster your career? There are the obvious answers of writing better reports and publications, presenting at conferences, and interacting more with your colleagues. Other answers should be tied into the other efforts in your career, such as networking and collaborating.

In using your writing skills, one warning reminder must be made in this new era of electronic communicating. Although e-mail is convenient, it does have its drawbacks. The written words do not carry any nuances or tones. Many meanings can be drawn from the same words. E-mail can be very impersonal. It makes being rude easy. In a face-to-face conversation or in a telephone call, you cannot ignore the other person unless you use tact. The impersonality of e-mail has a touch of anonymity and unconnectedness in it that can lead people to be more aggressive and less sensitive to others in ways that they would not be in other communication forms. E-mails are sometimes too convenient and familiar. There are no reminders inherent in them that you must reply to it.

3.1.2 Eloquence – Speaking Easily the First Time

Formal speaking is an inherent part of every good research chemist's career. If the person is in academia, then giving lectures to a class is a minimum. In government agencies and industry, formal presentations are given to managers and as part of internal technical meetings. In all three venues there is ample opportunity to give presentations at conferences. As a person progresses and gains more skills at speaking, it becomes easier and easier. With this in mind, the most difficult presentation must be the first formal one. For many researchers this is given even before the career starts, while the person is still in graduate school. I write this chapter with the emphasis towards that totally inexperienced speaker. The ideas and tips, however, are useable and useful to any speaker.

While attending a conference, I have often seen a young scientist nervously waiting for the time to give his or her first formal presentation. If the opportunity arises, I often try to give them a little more confidence and a few tips on speaking. My empathy arises both from remembering my first few talks and from my gratitude for fortunately having had some formal training in public speaking when I was in my teens. I realized that my good luck is not common and that few scientists ever have any formal training in speaking.

When preparing for a presentation, the first thing to do is to get a definition of the audience. The second thing is to choose the message you want to deliver to that audience. Scientists, in the course of their careers, may speak at one level to peers in a focused technical conference, to students in a classroom, to managers in a company or government agency, to more general listeners at a society meeting, to the public, and in other forums. Each group has a different level of technical background and makeup. The presentations to each must differ.

For most beginning scientists the audiences are other scientists in either a focused conference/symposium or at a society's meeting. In the first case, assumptions can be made about the level of background knowledge being higher and the interest in experimental details and results being keener than in the latter. For example, in a symposium on new insights on the mechanisms of noble-metal-catalyzed reactions, the speaker will not review as much as might be needed in an organic chemistry session of a national societal meeting. The expected level of expertise of each audience is different. In the first, everyone will understand to some degree the chemistry of platinum, palladium, iridium, and rhodium as catalysts. Why else would they attend? In the other, many will have passing knowledge, but some might not even be organic chemists at all.

Each speaker must choose what notes or prompts are built within a presentation. These let the speaker know what key points to make in each segment and when to smoothly move on to the next segment. I have found that a few index cards with brief notes of key points are useful. If the speaker loses place, it can readily be found again from the simple notes. Additionally, if a point is overlooked this can more naturally be handled in this loose format. Practicing the talk a few times to become familiar with the visuals and the points to be made is a good idea. Dr. Richard Mathies points out that one review or practice as soon before the talk as is reasonably possible both refreshed the points and clears the mind of distraction.

Conversely, using a highly scripted text does not work well for most people. This requires a lot of time and effort for memorization. The tone of delivery does not seem natural. It often is dull and flat in tone, which does not keep the audience's attention. It lacks any hint of enthusiasm. If one's place is lost, it is very awkward to resume the presentation without fumbling around. An audience does not like reiteration of material only because the speaker needed to find the scripted flow again.

The almost unanimous consensus among the dozens of scientists who told me of their preparations for speaking is that the more scripted a talk is, the worse it sounds. Memorized talks or those where every word is scripted and read aloud, come across as unenthusiastic, less knowledgeable, and less interesting. Speakers doing this often repeat themselves and falter. They fumble along from point to point and lose their audiences attention.

One tip is to start building the presentation with the most comfortable material as possible. This is often the background history of the research, the experimental description, and acknowledging other coworkers. The title visual helps introduce colleagues and a little background. A specific visual or two with background highlights are next and should be easily and comfortably given by the speaker and received by the audience. This is old stuff, often of common knowledge or at least it can be presented as so. The speaker should know the equipment and reagents used and cover them well. These segments often are several minutes long. This is sufficient time for the speaker to get started and well along in the allotted time.

As far as the overall content, the presentation is a sequential story that describes the research. Background, experimental details, results, conclusion, and possible future plans are a good sequence. Intersperse these with acknowledgements where they fit well. Do not worry if certain key points seem redundant. Most listeners will not catch every detail, no matter how well you present them and how hard the listeners pay attention (even those who are attentive and take notes will not hear all of the material – they get distracted by their own thoughts and recollections or they may not understand every new thing). Repeating the key points, particularly with a summary section, can have more impact. A simple figure with a few bulleted items will reinforce the message described earlier.

The visuals can either be chosen first to frame the story or the story – the message that will be given – can be thought of and the visuals to convey it then follow. The choice is whichever is most comfortable for the speaker. Visuals should be readable and simple. This means no small fonts and only enough items and details for a one- or two-minute description. If something is more complex and involved, use more than one visual, such as a general one with others showing specific portions. This works especially well for describing experimental setups or complex results.

There are many myths and fallacies that people have about speaking formally at a conference. One that I have found is that the audience sits on the edges of their chairs waiting for a technical misstep on the speaker's part. The young speaker almost quakes in anticipation of being caught in such a slipup. In fact, very few people in the audience ever try to show off in this fashion. The audience at a session of a technical meeting is generally genuinely interested in the topic and wants to learn about the innovations. This is particularly true at larger meetings where people have many concurrent sessions to choose from. They are willing to take the time to listen to a particular talk. They are an inherently sympathetic audience. They are waiting for the science, the technical details and description. They often neither care nor notice much about good speaking styles or polish.

A second fallacy, that students in particular have, is that the audience members know much more about the topic of the talk than the young, inexperienced speaker. If this were true, then the research is not research. It is not really new. If it is new, then those aspects are novel and the speaker will know much more of what was done than anyone else in the audience ever could.

A corollary to this is that if you are asked a question and do not know the answer or do not remember it, and then you can say that. Although people expect you to know more than they do when you are discussing your own research, they do not expect you to know everything as far as conclusions, interpretation, theories, or possible future work. When someone makes what seems to be a valid alternative explanation, recognize it by saying something like "That is a good observation. I'll need to think over that interpretation and its consequences." This approach both acknowledges a possibly good idea and that you are open minded.

Certain aspects of the fear of speaking are not related to technical issues, but are true of any speaking to groups and in a formal setting. Most people fear being the focus of attention, with a large number of others listening carefully to every word that is said, and all eyes focused on the speaker. Most people have this fear of speaking to an audience in a formal setting, a specific aspect of the stage fright that performers often speak of. Even after dozens and dozens of presentations, it is normal to feel a little anxiety before one speaks. The key is to not let that anxiety get to be so large that it impedes the speaking. How can one do this?

First, the speaker must feel comfortable with the basic premise of telling a story that describes the research. This should inherently be easy since the speaker is describing the story of her or his own research work. Many trainers in public speaking point out that this fear of speaking is mainly a psychological barrier rooted in the individual. If someone can describe their work informally to an individual or a small group of friends or colleagues, then the bulk of the basic mechanics of a formal presentation are already in place. These are an organized, sequential, and concise explanation, with further details given as needed. The major difference is usually in the formality of the visuals. An informal explanation relies on sketches on a notepad or chalkboard as needed, while these are replaced by viewgraphs or slides in a formal setting. The telling of the stories are in essence the same informally or formally.

One technique (or trick) that I was told of years ago is to speak to different individuals scattered throughout the audience. It is much more natural for a person to speak to another individual than to a large group. In order not to appear as if one were really focused on one audience member, shift the focus every minute or two to another person in a different part of the room.

Another trick is to have someone else watch you talking. If this is not possible use a mirror to watch yourself. Observe only for mannerisms and tones of voice. A varying tone of voice, reflecting humor, curiosity, amazement, and other emotions, keeps an audiences' attention. Moderate use of gestures, such as enumerating while sequentially using ones fingers, can be very effective in creating the image of a polished presentation.

A third pointer is to as calmly as possible apologize for any mistakes and deal with them. If a speaker realizes that a key point was not given in an earlier part of the talk, then a reference to that oversight and stating that point will both flow smoothly and be accepted by the audience. Every speaker has on occasion made an error in presentation and had to recover from it. Do not let this come across as anything more than a minor slip, correct it, and continue your presentation. Poise is noticed by an audience and if you project it you can keep the audience focused on your material, not your delivery.

It is unnatural for most people to talk while motionless. This, however, is the stance some take when giving a presentation. Using gestures with the motion of the hands, the arms, and different postures are one's normal way of speaking. Whether you speak behind a lectern or in the open (not behind a podium or lectern), your hands can be used to emphasize points. If you are speaking in the open you can also vary your stance and even casually move slightly as you speak.

Changes in tone and inflection are all a part of natural speaking. Pay attention to your own and make a listing. Look at this list and imagine if these seem normal ones for yourself and if you have seen speakers use them. If so, try to establish those in your talk. If you point at the screen, count off points made, or ask rhetorical questions in your everyday speech, these can fit into your presentations. Gestures and moving can be very effective accents in a presentation. A little caution, however, is not to do much more than what is customary in daily talking. These can be distracting if overly done.

If one is speaking behind a lectern, remember that it is an aid to hold one's notes and also to hold the slide control or pointer. It is not meant to be a support for the speaker to grasp, lean on, or be held up by. Assume normal speaking postures when you speak behind a lectern.

One's graduate school research colleagues are often a much tougher audience than the ones at conferences. The colleagues are often very familiar with the apparatus and techniques being used and of the problem that is being studied. Their questions come from the strength of knowledge. At many conferences, questions may be of a fundamental nature and readily answered. They also may be only out of curiosity. Even more commonly, they are just asked so that there is a question or two asked of the speaker. These types of questions are all easily answered because they do not come from strength of knowledge.

When discussing his preparations for a talk, one scientist pointed out that listeners vary. They fall into three groups, auditory, visual, and tactile- kinesthetic. The first two are much larger than the last one. They cue and learn from different sensory input. Auditory people listen to the speaker's words. Visual people learn most from the slides, overheads, and other graphical presentation materials, and that last group remembers best something real to their senses of touch or smell.

Visuals, whether slides, overheads, electronic projected media, or hand-drawn figures must be chosen to tell the story. The presentation connects these and gives details. In today's computer-oriented presentations, some people overdo the visual effects with motion pictures, three-dimentional graphics, and other flashy features without looking at the material to see if these add or distract. Content of the visuals is the first criterion.

In order to give the full message to everyone, you must prepare the presentation to say, the visuals to show, and if possible something tangible for the last group (such as passing around a rock if you are speaking of mineral analysis or a jar containing leaves of the plant whose natural products are your synthesis target).

Too often a speaker will prepare for groups one or two, but not both. Good speaking is complimented by good visuals and vice versa. Take the time to work on good visuals. This is often much better to do before you prepare for the talk, as each visual can be used to structure its part of the talk and also be used as a prompting tool.

What is the weakest point for many in the audience in a formal presentation? The answer from most scientists, especially those of us who have chaired sessions, is keeping the talk short enough to keep on schedule. Many in the audience have chosen to come to a specific session, or even a specific talk or two. Many times they want to go from one session to another in which talks are juxtaposed. Racing from one session to another is one of the realities of many of today's meeting, but it can only happen if sessions keep to their schedules.

How can a speaker plan to keep on schedule? A common rule of thumb is one visual aid for every two or so minutes of the presentation. This approximation depends on the complexity of the visual and explanation that goes with it. A title or acknowledgement slide is usually brief. One showing an experimental scheme or apparatus may take longer than the average. As another hint, if the schedule is running late and you can easily shorten your talk, the audience and the session chair will both appreciate that you helped get things back on schedule.

Is speaking in front of a large audience simple and easy? Not really, even for many experienced speakers. But if one understands that many of the common fears are unfounded, then this situation should be much, much less fearful. Preparation and a good frame of mind go a long way to creating a good first presentation.

3.1.3   Writing a Paper – The Basic Mechanics Help

Writing is a necessary skill in whatever venue a person works. In some, such as academia, it is a key differentiator between succeeding and only getting by. Grant applications and publications cannot be avoided in academia. They are almost as requisite in government laboratories. This skill, however, is not inherent in most people. Many professors cited it as the weakest skill in their graduate students, the one requiring the most training.

Every opportunity to improve writing skills should be taken. If there are formal courses offered, enroll in them. Look at the various books listed in the bibliography. Write abstracts for every presentation you present. Write detailed notebooks. The more you write, the more skilled you will become.

When the task of writing a report or manuscript must be

done, some people write it from start to finish, in the order planned. They write each sentence and paragraph sequentially. Each difficulty in wording and expression holds up further progress until it is surmounted. This approach is one that is not necessary for everyone. It is slow and ineffective for many writers. Find whatever writing approach that is most comfortable to you.

Sections of reports or manuscripts differ in content. This means that they can be written separately with any minor revision written afterwards to bring them together as the completed work. This approach works for many writers since they can write in any section in which a new thought is made.

In the past, when writing by hand or typing had to be done, this approach was as good as any. With the features of today's word processing programs on personal computers, it is an anachronism. Sections of a manuscript can be written piecemeal as the various ideas are thought of. Each idea is written and saved. When the text seems complete, reading for disjointed section or redundancies must be done. The cut-and-paste feature can easily help move sections so that the themes flow evenly and thoughts presented are not disjointed. Redundant statements can be deleted, leaving a statement in its most appropriate place.

An alternative to this approach is to create an outline or set of notes. Each idea is added as the form of the written work is planned. As this proceeds, the details get smaller and smaller until the finished outline contains one line for each planned sentence. The various groupings in the outline correspond to sections and individual paragraphs. Iterations of more and more detailed outlines will eventually end up as the complete framework for a manuscript. The topical headings in the outline are converted to complete sentences with more details.

Another approach was described to me by an imminent organic chemist. His very first step is to think of the target venue, which journal the paper should be submitted to. There is a quick review of the guidelines for authors for that journal, including formats for footnotes, tables, figures, and references. He then relies on the visual nature of the work being described and on the fact that some authors are better with conceptualized ideas. The first step he uses is to think of the figures that need to be shown to make all of the important points. The captions and text describing each figure give some of the explanation. This leads into writing the connecting sequences and other details. The results and discussion are often the first part written, with the experimental conditions being next. The introduction is often the next to last part written, with only the abstract being done later.

There are now features in word processing programs that track superscripts or subscripts. These features are perfect for numbering references. This saves having to renumber as newly found references and footnotes are added. Some of the features in these programs are obviously easy to use, such as spelling and grammar checking. The writer, however, must remember that these are not as sophisticated as a proofreader. Words mistyped into other words will not be caught. The grammar errors that are found may help non-native speakers write better by catching noun-verb tense and number mismatches. They will not work as well in scientific writing because so many words are not in the program's dictionary and the grammar of passive voice may throw the program off. Mistyping a chemical name will give the same lack of recognition as typing it correctly. A writer might get in the bad habit of quickly telling the program to ignore chemical names that pop up without looking to see if they are typed correctly.

One researcher strongly emphasized the different audiences that a manuscript might be aimed at. He then described the differences in writing a technical paper, a review for a topical technical journal, and for a general technical journal. His first step is asking the question "What message do I want the reader to get?' The introduction of a technical paper, for example, describes why the current research is important. For a review, it describes why that area of research is important.

Writing an Eye-Catching, High-Impact Technical Paper

Most chemists think that the only criterion for an excellently received manuscript, one in which the reviewers find few revisions and think the work described is excellent, resides only in the science. As both an interested chemist reading published papers as well as having authored or coauthored well over a hundred peer-reviewed papers and as a reviewer for more than a score of journals through the past decade, I can say that this is not so. Writing and describing the work in a clear, concise, but interesting manner, moves a paper from being good to being exceptional.

If the same experimental work is described by two different writers, very different levels of manuscripts can result. Great science when written poorly can be a good paper or alternatively, average science when written well can be a good paper. On the other hand, great science when written well can be an excellent paper and average science when written poorly can be the readily rejected paper.

First of all, the "mechanics" or "structure" of a paper must be done appropriately. This means that the information given and its descriptions of background, experiments, data, and results conform to the asked-for standards. There are some useful resources for writers that make creating a good manuscript easier. One is, of course, carefully reading the instructions for authors that each journal has. These give broad descriptions of what each journal requires in format and arrangement. Different journals, for example, have specific formats and rules for numbering references. Some allow or even require citing unpublished data, while others do not. Some cluster similar references under one number, while others require a separate one for each cited paper. Reviewers are people and most people, once their eyes catch errors such as poor grammar, misspellings, or misused terms or acronyms, will look for more of these types of errors.

Another resource is the books such as the American Chemical Society duo of publications "The ACS Style Guide" and "The Handbook for Authors". Both give accepted uses of punctuation marks, acronyms, SI units, and other details. They also contain detailed listings of printers' proof marks which are very useful in correcting the proofs of manuscripts. Although these two books are aimed at writers of manuscripts destined for submission to the journals published by the ACS, many other journals use these guidelines or a vast proportion of them. These types of guides ensure that a manuscript technically conforms to the standard practice for the targeted journal. Without this conformity, a manuscript may require severe revision because it is not in suitable form.

Sometimes when reviewing a manuscript, it seems that

authors even skip the basic steps like using the "spell check" features in all word processing programs or confirming text referencing to the numbers in the list of references. It may be difficult to read a manuscript for such errors after writing and rewriting it several times. The brain naturally scans through and reads the content after so many readings of the same sentences.

An author or group of coauthors should give the manuscript to someone less familiar to it so that such easily overlooked mistakes can be caught before submission. Reviewers often think an apparently sloppily prepared manuscript must reflect sloppily done science. This may not be the case, but human nature sometimes creates that connection. It is much better to avoid that possibility altogether.

If the basics are dealt with by following the manuscript guidelines, then what adds to a manuscript's readability? What elements go into writing an excellent paper? These do vary depending on the research described and on the publication.

The first rule of thumb for authors must always be "Write for the readers". Papers are chronicles of work to be read by others. The writers are intimately familiar with the work and concepts. Efforts must be made to look at the manuscript from a more distant perspective. Readers, even experts in the particular field, may not be as familiar as the writers are. This is not to say that the writers must cover everything as if novices are reading it or from a basic viewpoint, but care must be taken to avoid assuming too high of a level of familiarity. Acronyms and other abbreviations denoting a phrase should always be defined at the first place in the text in which the phrase is used.

I will first describe the content and approaches to a manuscript that reports experimental work. It always helps to have a good introduction. A thorough description and referencing of the background work that led up to the current research starts a manuscript off well, the idea that the work being reported is part of the continuous evolution of knowledge. This creates both a stronger feeling of interest for the following sections and a good sense about the work's quality. The importance of the introduction is highlighted by the fact that many authors told of writing it last, not only because it leads into the experiments, results, and conclusion. It must set the reader up to be receptive to the later sections.

First impressions are important. Most research articles contain an introduction of this sort, but there is a variety in the quality and effectiveness. This introduction must build up in a logical progression to the new work. What led up to the current work? Why was the earlier work both key and limited in not explaining everything? Thus, the current research fits a strong need to answer previously unanswered and important questions.

A well-written introduction also gives the readers practice in reading the papers ideas and rhythms. Writers, even in technical papers, have styles. Some write in short sentences. Others write in long ones with many subjunctive clauses. Some use a simple vocabulary. Others use less often heard words to be more precise or expressive. A reader subconsciously gathers these differences as he or she starts reading a paper. The more difficult sections to read and understand are the later ones of results and conclusions. By the time a reader gets to those sections, a well written manuscript will be easily read and understood.

The introduction should also contain a section describing why this work was done and its importance. This is often omitted because the authors assume that it must be inherently obvious. Some readers may understand the implied importance, but many will not. It does not hurt a paper to have a few paragraphs on what makes the work noteworthy. These statements may also catch the attention of those who focus more on applications and potential rather than techniques.

The descriptions of the experiment must be clear enough so that they can be readily replicated. This is much better if it is done without delving through several other references to get the gist. One of my quibbles with some manuscripts is too much brevity in the experimental section or relying on numerous references to describe and explain the methodologies, with the experimental setup being briefly covered and the sources of chemicals and apparatus being the only other things described. This problem of relying on references to describe the work is compounded even more if those references are in less accessible journals or in languages other than English.

The experimental section is often the easiest section for many people to write. It is a logical first section to get the writing underway. The various tasks and things used to do them are well defined and involve little explanation or interpretation. A writer must remember though that the readers are never as familiar with the experiments as she or he is. Is there enough detail to replicate the work elsewhere?

The results and conclusions must explain the interpretation and applications covered by the data, plus the limitations and further work that needs doing. Many papers contain results and tables or graphs of data without much discussion of how good the approach may be over earlier alternatives, repeatability or reproducibility, or interferences and any limitations of the method. These areas are often very important ones to potential users. Having to go through further development or trial-and-error in implementing what appeared to be a valid method in the publication can often lead to rejection of use of the method.

A writer must take care in spelling, syntax, and grammar. If the principle author cannot easily do this, then the other coauthors must take on that responsibility. If the writers are not fluent English speakers and writers, then they must find and rely on someone who is. This person must look carefully through the manuscript before submission. This is especially true when English is the person's second or third language. This is not only to make the work understandable and clear to readers, but this helps in the review process. Reviewers are human underneath their technical competencies. Consciously or subconsciously they read through manuscripts with biases, good or bad. If they struggle to understand a paper, a negative image of it is created that may tinge other aspects of the review. They may become even more stringent in their criteria for approval.

The selection of illustrations, tables, and graphs are not just to present the data and apparatus described. They must be effective in this. Too many or too few illustrations can greatly weaken a paper. On one hand, too few leave details as ambiguous. On the other, too many can be redundant, distracting, and divert the attention of the reader from the main points.

Many people are more visual in the way they assimilate information. To them, a few good illustrations make much more impact than pages of well-written text. Thus, illustrations are an important tool to clarify and augment the text. They are not just a reiteration of text. Choosing a good series of illustrations can make a good framework for the manuscript. The text explains the illustrations in more detail when this approach is used.

One area that many authors overlook or give only minor attention to is the acknowledgements. Although the people and organizations included there do not affect the readability of the manuscript, many authors have made grievous errors of omission in their acknowledgements. Leaving out someone who has aided through key ideas, review of the manuscript, donation of equipment, samples, or rare chemicals can lead to injured feelings. Subsequent assistance may not be so readily forthcoming.

Collaborating on Jointly Written Work

The dynamics of writing are similar, but not identical, if the preparation is a group effort. Often one writer, the most involved in the work or most senior researcher, will write the first draft. Certain sections may be written by the others if they were more involved and merged into that draft version.

I have run into a difference in styles in collaborative writing which points out that there must be a plan for the manuscript. This plan should have what each person's responsibilities and how the various parts will be joined together. I am a flexible person, seeing that many routes may be possible to get to the goal need. In one of my projects, this was a slight difficulty because my written sections were all draft versions in my own mind, open to any suggested editing. Others on the team were more the kinds of writers who do the rewriting themselves and expect what is shared to be a near-final version. So I had to modify my style, making editing it and re-writing it part of my draft version. It was difficult, as I did not know what the others expected other than it be near complete.

The Ethics of Authorship

One aspect of writing manuscripts was highlighted in an article in Chemical and Engineering News (February 28, 2003, page 31). It describes a report in an organic chemistry journal in which there appears to be two rather dubious practices in manuscript writing. One is not giving a full listing of valid references by omitting research from competing researchers.

106

The other is not referencing one's own work published in other journals because there is a degree of "self plagiarism" going on, meaning publication of similar research in different journals in order to increase the number of publications. Cross-referencing might bring this practice into the open. Most journals have explicit policies against this that submitting authors are agreeing to. Research that is reported must not be already published. The only exception to this sort of publishing might be for review articles which are targeting a different audience. In that case, even, the manuscripts must be rewritten and re-referenced to accommodate each audience. Submitting a very similar manuscript as a review article to two places is unethical.

The Reviewing Process and Later

Submission of a manuscript is often nowhere near the last step in publication. Journals use peer reviews to assess the manuscripts they receive. The reviewers are selected for their expertise by the editors, but it must be remembered that reviewers are not omniscient. They sometimes make mistakes or put forward opinions that may not be correct. That being said, however, most reviewers raise valid questions about the references that are and are not cited, the research described, the experimental design and execution, and the interpretation of results. Their comments are aimed at both maintaining a high quality for the journal and to aid authors in reaching that level if it is possible. Authors can suggest reviewers in their submission letter. Some journals even ask for these nominations. Suggesting reviewers must be done ethically. The scientists named must not be anyone who might favorably review the manuscript because of personal friendship or an on-going collaboration on another project.

Good reviews are easy to deal with. Praise is readily accepted. Minor changes, especially in grammar, spelling, and other non-technical areas, are seldom an issue for authors to change to the suggestions of reviewers. There is little time or effort to making the needed changes in the bulk of cases when reviewers like a manuscript.

Negative reviews can be both ego-deflating and difficult. The first emotional responses that occur when reading criticism must be set aside. Then you must read the reviews to see if there are valid points. This does not mean those that might be totally true, but also all that may be partially to a degree or true on certain points – if not on all of the ones given. Are there really too many figures? The reviewer may say so and suggest eliminating certain ones. The authors must look at each and decide if they really are absolutely needed to tell the story. Are the figures redundant with the text or does a smaller table or the text do that as well?

You must be honest with yourself. Is the manuscript or work really bad? This is especially true if it gets rejected. Rejection is hard to accept for anyone on anything, but that does not mean it is not warranted. I once had one rejected and the editor noted that it was not up to my usual standards. I looked at it and had to agree. The work was not as thorough or innovative as I had thought. I had done the work in a hurry to meet a submission deadline for a conference. I made that deadline, but the quality suffered for my haste. If your paper is rejected, let a few moments pass in order to calm down. Then as unemotionally as possible, read through the reviewers' reasons. You may feel the need to unleash your natural tendency to refute the reviewers and start to list counter-arguments to their points, but wait to do this until you have looked at the reviewers' points with balance. Then read through the comments again and make statements supporting the reviewers' arguments. See if the supporting statements or the counter-arguments seem more valid. Then decide if the paper is really good or not.

Sometimes different reviewers come to different opinions on the same manuscript. The editor is supposed to notice this and reconcile the viewpoints before sending the reviews to the authors. This may not always happen. The authors, however, must take differences of opinion seriously. They should not just point out to the editor that there is agreement of the favorable reviewers, that one reviewer disagrees, and thus leave those points of the manuscript intact.

In my own research work, I had a manuscript that described the synthesis of several new related compounds and the trends seen in their chromatographic and spectroscopic behaviors. There were three reviewers. Each made strong supportive statements, in great detail, in synthesis or chromatography or spectroscopy. They then each quibbled over points in the other two areas. My reply to the editor was that each was obviously expert in the first area, as shown by the detailed comments. Their expertise in the others clashed with each other. So, he needed to decide if the overall reviews were accepting with little changes or rejecting with great changes. Fortunately, he chose the first.

A touch of tact is needed when sending back comments in reply to reviews, even when that was not the case within the reviews. The reviewers generally hold more power. They decide if a manuscript is accepted. Rarely is there research that is so innovative, so compelling, and of such impact that a journal will give the authors the leeway in a dispute.

Acquiescing to a reviewer's comments is a balance between how much you want the paper published in that particular journal and how strongly you disagree with the changes put forward. Most authors will submit a manuscript to another journal if they find the changes objectionable.

In the extreme case where the negative reviews cannot be surmounted or appear to be unfair, authors can ask the editor to act as a referee or for him or her to send the manuscript to other people in the field to act as referees. The referee's role is often limited only to deciding if the authors or the reviewer's opinions should be accepted.

In no situation should an author agree to reviewers' comments, submit an altered manuscript which is accepted, and then revert to the original version in the proofing process. This is a breach of faith. Printers and proof editors are not tasked with technical merit. This type of behavior will only be noticed if the changes in proof are substantially different than the approved version. Reviewers and editors must ensure that this does not happen. If it does, then a reviewer who observes this should notify the editor. Editors should then sanction the authors along whatever ethical guidelines the journal may adhere to, including a further ban on publishing in that journal.

The final step in getting a paper published is proofing the typeset text. Proofing of the final typeset material is also important, but it can be tedious. It is difficult to read a manuscript for printing errors, proper punctuation, or other things looked for in page proofs. This is especially true after writing and rewriting the manuscript dozens of times. The mind and eye are used to reading for content and the sense of the words. This makes the reading of proof ineffective because many errors would get overlooked. One such example would be the double wording where a sentence contains the same word at the end of one line of text and also at the beginning of the next line. The mind reads as if there is only one.

One colleague surmounts this by looking at the text slowly, sentence by sentence, but starting at the end and going backwards. The context and meaning are lost from the mind. The focus is on spelling, punctuation, and other such errors.

This approach may be too tedious for most in order to be effective. My own approach is to first read each sentence word by word, then to read each sentence as a whole for grammar such as verb tense, for noun-verb matching, and sentence fragments. I keep the ACS Style Guide at hand to insert the appropriate proof marks. One area of particularly close inspection is the reference section. Referring people to the right sources is one aim, but another that I have found is to ensure the spelling of the various authors' names. My observation is that nothing irritates colleagues more than not getting their names correct.

Completing the Circle – Doing Things When the Paper Is Published

This section does not deal with things related to writing and publishing a paper. It covers several things that are connected to other career aspects that derive from getting a paper published.

When your paper appears, add that information to your list of publications and to your curriculum vitae. This ensures both are up-to-date. A separate publication listing is both a redundant listing for insurance and is in a more comprehensive and chronological than the split version in a CV. It is always a good idea to keep a file folder with at least two paper copies of every one of your publications. Doing this makes sure that you have every publication readily at hand. In the instance where you might loan one copy to someone, you would still have another in hand for your own use. An electronic file system of this type may seem more modern and convenient, but data formats change. After a few years your stock of publications contains a lot of old meaningless gibberish because the older program files cannot be read.

When you receive reprints, add copies to your publications file. If the paper is one in which you did not receive reprints, then make two photocopies from the journal to serve the same purposes. This may not seem to be a major need, but as time passes the availability of copies from journals may change. You may loan out your copy of a book with a chapter by you in it. The borrower loses it or in the mix of loaning books out you forget who has what. I admit to this happening to me, so that I do not have all of the books that my chapters have appeared in, but I still have copies of those chapters.

Set aside several reprints for each other coauthor. For those coauthors not at hand, send those with a note acknowledging their participation and your thanks for it. For those coauthors where you work, pass out reprints personally with the same message. Do the same for anyone who was acknowledged in the paper. Write down a list of people in your network who would benefit from receiving a reprint. Set aside copies for that so that they can be sent. When you do this, add a note highlighting any parts that might be of particular interest in their research. Other people appreciate this. It helps them in their work and often gains you a future citation when they prepare a manuscript. Do not ever assume that everyone in your network must have seen your paper. They are as busy as you are and might not have been able to.

You will receive reprint requests. These are often sent by scientists who do not have access to the particular journal you published in. They see a citation or a listing in a table of contents service. Sending them reprints may seem a tedious chore, but think of yourself being in that situation. You would need and appreciate a reprint, so you request one.

3.1.4 Writing with Style

Once you understand the structure of a manuscript and the mechanics of writing one, you can then move on to making those things easier. Practice is one way, by writing more and more you will get better and better. Another way which is less used but is also effective is to write in a style that reflects your thinking and speaking. A more natural writing style is easier since you already put your ideas into words in that fashion when you communicate orally.

Even those comfortable communications styles differ depending on the audience you are sending the message to. The same story of your research with a close colleague who you chat with every morning over mugs of coffee would be different than the update you give to your supervisor. With any of these, however, you use words, figures of speech, analogies, humor and other devices to get your points across.

A similar aspect of writing is that there are a variety of writing styles fitting your personality and the audience. For some types of writing, such as for reports, a factual delivery that is often terse is the normal way. The simple verb tenses are preferred, often only in the past or present tense. If the report describes something or delivers factual information, then it is often in sentences such as "The salinity of the water changed with depth," or "The gross national product in 2000 was X Euros." If a report describes work done, it is often written in the first person. Almost always the active voice is used instead of the passive voice. There is little use of descriptive adverbs or adjectives. The use of similes, metaphors, and other figures of speech are almost never done.

Many technical publications, on the contrary, can contain more complex sentence structures. These papers describe much more complex experiments, theories, data observations, and interpretation. Participle phrases, subclauses, and passive voice are some things seen most commonly in technical papers that are not found in written reports. Still there is little use of descriptive adverbs and adjectives or figures of speech. Most technical papers sound a little stilted because passive voice rather than active voice is used. Normal speaking styles, the communication most people are familiar with, is in active voice.

Why Are the Accepted Writing Styles in Technical Publications so Narrow?

"Once upon a time…." I started this section in this fashion because I plan to touch on the writing styles of technical papers. Any scientific paper that began in like that or in any other humorous or whimsical tone is almost immediately received negatively by reviewers. If the paper does emerge through the review process with this touch of humor, many readers will also react negatively. A large portion of the scientific community has very little room for humor and other "literary" inclinations in technical publications.

This attitude is in stark contrast to the corresponding one for presentations at symposia and conferences. Successful speakers weave humor and more descriptive, and often much less commonly used, words into their talks. In fact many listeners of a totally factual presentation of the same tone as that in the eventual technical paper would find themselves losing attention and their thoughts drifting off to other things. My contention and purpose with this section is to state that writing styles for technical papers do not have to be totally devoid of the writer's personality.

The contents of the technical paper must be much more detailed than the description given in a fifteen or twenty minute talk. The giving of a much more detailed description, however, cannot be the reason why we expect papers to be so much more straightforward and dry than the analogous talk.

Why is there this difference in expectations between the verbal and written descriptions of the same experimental work? There must be many answers, but some of the most commonly heard ones arise from what we have been taught to expect from each. For a technical scientific paper these are that everything is told in the passive voice, with a beginning background and history, the experimental details of equipment, reagents, reactions, etc., followed by results, interpretation, and conclusions. For a technical presentation the active voice is often used rather than the passive. The various descriptive sections are similar in each. The stylistic freedom allowed in a presentation, however, is much wider. The little touches of personality and humor given in a talk can be readily transferable to the written version, but they are most often not.

The overall structures used in a paper are useful and lead to a logical telling of the experimental story, aimed at anyone who might want to replicate the work. These aims are gained through the experimental sections describing reagents and apparatus and in the results section. Is this factual telling the only aim in an author's writing of or in a reader's looking at, a paper? I think readers look for and writers should include more of the personality and the thinking behind the experimental planning and in the conclusions reached. These are often not given very well in papers.

I had a colleague who once submitted a manuscript to one of the leading chromatography journals. The science was good and well-described along the lines mentioned above about covering the experiments and results. My friend, however, is an eloquent person and uses a lot of metaphor and colloquialisms in his language. He started the background section of this manuscript with the phrase "Such-and-such technique has fallen on hard times." He then detailed why this once widely used technique had become less favored and less used.

When he received the referees' comments for his manuscript, there was little change needed for the technical information or for his conclusions. One referee, however, said that the paper needed major revision to remove the "Dickinsian prose". The tone of that first sentence of the manuscript was too literary for that reviewer. His reading of the manuscript then included noticing any other literary touches. Although this became an on-going joke between my colleague and I, I know that he was bothered that differences in the stylistic theories of scientific writing changed his words from that describing his own work into something that seemed less reflective of him.

Do we expect every published paper to be so identically written that a description of research work would be essentially the same no matter which researcher did the writing? This sounds like such a restrictive attitude that very dry, factual-only manuscripts would be the only result. We, of course, do not want the writing to become so stylish that much of it serves only artistic and aesthetic purposes. After all, scientific literature is a fact-based description of research and observations. There can be a balance of the factual with some stylistic elements.

Scientific manuscripts are written by people who are scientists. The research is done by people. To expect writers to totally avoid or remove any aspects of their personalities seems to be a very harsh and stark goal. A touch of humor or a few words chosen for style should be commended if they add to the paper. They should only be lamented if they detract from the describing of the science.

Reviewers might include a readability factor in their judgments. If done now, this is often only a cursory assessment of the quality of English used if the writers are not primary-language English speakers. Readability should look at the avoidance of trite or turgid phrases and redundant use of words throughout the text, as well as the positive use of interesting words and phrases that accurately capture the message that they want to give the readers.

While very "flowery" language may be distracting to a reader, the use of a wider range of adverbs and adjective will help liven a paper and make it more readable. Careful use of figures of speech can, too. Thus, instead of "the temperature increased very rapidly", "a precipitous rise" says the same thing. It also is different than what is in most papers and makes the reading more varied and enjoyable.

This must be done carefully, however, because idioms and colloquialisms are rife in our everyday speaking. By these, I mean those phrases of very specific and localized usage that all readers may not understand. In our global society, an English speaker from Scotland may use a phrase that is obscure to even someone from Wales or England. This is even more befuddling to readers in the various regions of the United States, Canada, Australia, New Zealand, and other nations where English is the primary language. The readers who speak a language other than English as their primary language would have no idea of what the writer is trying to convey. These phrases of only limited geographical understanding should not be used.

Humor is present in much of our research. It sometimes jumps out in the names given to things, such as NOSY and COSY in NMR or buckminsterfullerene as the sixty-carbon cluster. Scientists are generally as good as other people in telling and appreciating a good joke or pun. We, however, very rarely do this in our writings. It is another touch of individuality that should not be overlooked if the opportunity to use it is there.

One of my own insights into the skills in and variations of writing has been found through the creation of the series of essays that led to this book and, of course, the writing of it as well. In them, I have found that I am using many of the common descriptive words that I also use when I speak. This comes very naturally since these essays were being written to convey my thoughts and were the way that I would express the same thoughts verbally.

If you look through these pages, you will find many words of this sort. Ones that you read, know the meaning of without thinking, accept, and which create thoughts and images. Look back through the book at what you have read and you will realize that you did this. We understand many more words than what we commonly use. Using a wider vocabulary both conveys the exact meaning the writer wants and creates a style that reflects that person (did you stop at reading the "convey" and think that it is one of those lesser-used words?).

I also have tended to use many more common figures of speech in these essays. On the contrary, I do not as often do either of these in my technical writings. Why do I get more bland and hackneyed when I write a technical paper? I, too, subconsciously accept all those unwritten and unneeded rules that stifle my normal way of communicating.

To look at a different perspective, gives another good reason to accept papers with a broader range of writing styles. Reading scientific publications can be very tiring. There is concentration needed to understand and absorb the facts, data, and results. If one writes more readably, in consideration of the readers, the use of simile, metaphor, selective use of idioms and colloquialisms, and humor should all be some of the tools used. After all, one of the very human aspects in reading a journal is that it takes time and energy to do it thoroughly. If a reader is very diligent, several journals are regularly scanned and read. Lots of time and effort go into keeping up with the literature. Good writing makes this an easier task. Poor or dry writing makes this more like drudgery.

Writers can use the introduction to tell the tale of why this research area is so exciting, how the history of the research in this area has transpired, or why previous work has missed the mark. They can use the discussion to speculate, to describe future wished-for experiments. These add character and flavor to the research that reflects the personal emotions we all feel about our own research. When we read these portions of a paper, we should be able to feel some of that enthusiasm or contemplation.

A published paper does not have to be flashy. It does not have to be splashy. It does not have to be fine literature. It does not have to good comedy. It can, however, be human, containing touches reflecting each of us as people in addition to our being scientists. All of these can be done with some effort and more acceptance by reviewers and readers, without compromising technical quality.

3.1.5   Listening

When people think of communications skills two areas come into thought, speaking and writing. There is a third that is also important. The most often overlooked communications skill is that of listening. Whether in the context of hearing a presentation, discussing a project with a colleague, in discussing issues with one's supervisor (or with others that one supervises), or in numerous other contexts, the skills in listening are needed. To highlight why this is so, you must only look at communications through instant messaging. The nuances of speech are not present, but people still use their familiar speaking patterns and words. Misunderstanding happens very often because enthusiasm reads the same as sarcasm; a joke may look like a threat, and so on.

Since verbal communication involves an exchange between at least two people, each is thinking of thoughts to convey, how to express those words, and listening to the other for comprehension and the other's thoughts. It is continuous and complex enough to be imperfect. A moment's diversion in order to ponder a thought can lead to a loss of attention to the other person's speaking.

Why Are Listening Skills Valuable?

The aim of communication is to transfer thoughts from one person to another. This involves both the sending and receiving of the thoughts. A highly skilled speaker cannot communicate if his audience is unreceptive, misunderstanding, or inattentive. The ideas are transferred only when the audience is open, comprehending, and listening. In fact there can be little comprehension if the audience is not listening.

What Are the Skills Used in Listening?

One skill in listening is paying complete attention to the other person while he or she is speaking. "Patience is a virtue" is an old platitude. It is very valuable in listening. When there is supposed to be dialogue, people often listen for a period of time. Then they hear something the other person says and are diverted to their own thoughts and to the comments that they want to say in response. This stops the listening process or changes it into a selective process.

Selective listening can either be listening when one's mind is not thinking of responses or when one only listens for material to use in responding. In either case the speaker is only partially heard and the message that is to be conveyed is lost.

Speaking and listening, especially in a dialogue, are not races. Waiting and pausing to hear each other, to gather one's thoughts and the wording needed to describe them and to express them can all be done at a pace that gets the messages across. That is the most important thing, ensuring that a discussion succeeds because ideas and messages are exchanged. Why talk over anything if you do not plan for communication to happen?

So slowing down your speaking pattern and also allowing for the time to fully listen to the other person becomes a key to successful communication. Slowing down also allows the more careful selection of words and phrases to describe the ideas. Once real dialogue gets comfortable, speaking and listening compel each person to work harder at understanding the other.

Slowing a conversation down a little in order to fully listen is often difficult because most people are taught through personal experiences to speak rapidly. This may also involve interrupting someone else in mid-sentence to interject your own thoughts or comments. It is one of my own faults which I try constantly to get around by waiting. If I interrupt someone and remember in time, I pause before I speak and ask them to go on with their talking. This leads to an imbalance, as speaking is much more emphasized than listening and a dialogue is less complete.

Another aspect of this waiting approach, which is effective in a dialogue but even more useful in a group discussion, is to listen to the other comments and gather all the ideas and perspectives. Then when you choose to speak, you both have had time to consider the other viewpoints and gather your own. This allows modifications to be made to your original thoughts. If this becomes a habitual response in discussions, others will give more weigh to your comments because they will be more thorough. Waiting to speak last, or at least towards the end of the discussion, has the most impact.

This is not quite keeping to the adage "It is better to be thought of as dense (or clueless or stupid or naïve or any other negative), than to open one's mouth and proving it." There is a grain of truth in this, though. Speaking too soon may sometimes seem to show that you do not understand the discussion, its issues and complexities. Waiting may give you both the understanding and the situation to appear to be very knowledgeable.

Another skill in listening is to understand precisely the meanings being used by the speaker. This is the mirror-image of a speaker being able to express the thoughts to be sent in words that exactly capture the meaning. The most precise and carefully chosen words may express intricate ideas exactly, but they mean nothing if the listener cannot understand them. A limited vocabulary or reliance on trite and common phrases limits clear communication in both speakers and listeners.

For everyone who must communicate in writing and speaking, the dictionary and thesaurus are two resources that must be commonly used. Get in the habit of using them. Whenever you encounter an unknown or unclear word in your reading or listening, make a point of noting it for looking up. It is best to do this as soon as possible after seeing or hearing, as the connection between meanings and contexts are freshly made.

Make an effort to learn more words. This should be especially done after you hear words that turn out to exactly express an idea. An expansive vocabulary becomes a handy tool when you speak, write, or listen. It also has the additional benefit of often giving others a good impression of you.

When you listen and do not understand, ask for clarification. Do not listen on in hopes that the next statements will allow you to figure out what was meant by the context. Spend the time and energy to hear the message the other person is telling. It is even wise to sometimes ask in order to make certain that what you hear and understand is what the other person is trying to say. Reiterate in different language the message and see if there is agreement. Assuming understanding can be a serious mistake. This is sometimes even exacerbated when you say "I see" or I understand" and the message was not really understood. Close the loop by check back with the speaker. Make certain that the message being given is the message received.

Most people have another listening skill which they use only half consciously, that is listening for tone and inflections of meaning. The same phrase may be said jocularly or sarcastically, with skepticism or uncertainty, or with many other emotions behind them. We often use these clues in other aspects of our lives, but we cannot think that these are not present when we talk science.

Another facet of this that does not involve the sense of hearing, but which can be as important as hearing tones and inflections, is in reading the signs of body language. When the other person is quiet, it may mean that she or he is listening and thinking of what you are saying or distracted and thinking of the sport's score from the morning newspaper.

Body language is a sociological field of study in itself that requires too much description and background into the psychology for this book. Most people are aware of some, such as the crossing of arms signaling disagreement. There are even cultural differences in body language. A few resources are listed in the bibliography on this area.

3.2   Networking – Becoming an Integral Part of Your Field

Over the years I have been asked "How did you become interconnected with so many other chemists?" or "How did you get involved with this journal, committee, or society?" When I think of my answers to these sorts of questions, I have not replied with what most people might think. These connections of being part of the active scientific community are not a matter of selling one's self. This is the common, but wrong, assumption or perception. They arose because others knew of me through personal interactions. Networking has been an inherent part of my career even before I understood its advantages.

Networking is building a sense of connection and community among scientists that one knows. Unlike normal communal feelings, however, in this case one gets to choose and define who and what the community is since it is one's personal scientific network. Membership can be selective and exclusive, but very wide ranging under whatever criteria you chose.

Your network is defined by you and is your own resource. It becomes valuable professionally and personally. As time passes you will find commonalities and redundancies where those in your network have each other in their own networks. The networking becomes both cumulative and cascading. One member will refer you to other new members, who in turn refer you to even more.

When you interact for the first time with a person who you have been referred to by someone in your network, remember to mention that in your introduction. This should be true no matter what the context or format. This sets up the interaction in a beginning positive light. A mutual friend is a linkage, especially if that person recommended that you contact the other person. A corollary to this is to always let another know that they can contact one of the people in your network with you being the connection. The collegial mentality always opens the doors to new interactions easier.

In my case, the opportunities for collaboration and for being involved with journals or societies have most often been brought up by someone else asking me if I would be interested. In the case of collaborations and reviewing for journals I initiated the contact some of the time. The obvious next question that is asked about my getting these opportunities is "What things did you do to make yourself come into others' minds when they were in need of a person?" The first of several answers, of course, had to be good technical capabilities must have been shown since we were dealing in technical fields. How to gain those is the theme of some of the other essays in this series, but in brief it entails keeping up with the literature, doing innovative work, and presenting that work at conferences and in publications.

What non-technical factors lead to these opportunities? There are several answers that are more based on personality, which I can sum up in the phrase "It does not hurt if you are friendly and nice to others". This sounds simplistic and even too sugar-coated. But as with any social interactions, if you make a good human contact while being a technically successful chemist, then you will build a network of colleagues and friends. People will build strong links to those they enjoy interacting and working with.

Within each of us, underneath the veneer of the scientist is a person. Connecting with the person as well as the scientist is the basis of solid networking. The technical competence will create a scientific network, colleagues who have similar technical interests. This is the framework for building a personal network. The strengthening substance is filled in by the personal connections.

The first stumbling block to this is the same as in other social situations, how to break the ice. With scientists, however, there is the natural common interest of the science. Ask about the person's research and discuss it with the real interest you have as a chemist. You can expand your interactions by asking about the person's institution or company, their locale, or a colleague of theirs that you might know. There are many natural connecting topics. Listen to the person's comments and reply positively. Other topics often come up if you listen for them. Later, if you happen to see that person again, smile and say something to link again even if it is only "Nice to see you again. How is the conference going for you?" This recognition of someone only recently met can create a strong positive image.

Another thing to keep in mind during any interaction, whether by correspondence or directly, is that courtesy and politeness do help. Answering a query quickly, thanking and acknowledging someone for help, and many of the things we have learned to do in our daily lives should also be a part of our professional lives, too. It often helps as a reminder to try and imagine myself on the other end. If I send out a question to someone, would I rather receive a reply within a day or two or within a week or within a month, if at all? If I do something for someone, does the "thank you" they give make me feel better or does the one not said make me feel worse?

Corresponding is also a good way to build connections. If someone you are acquainted with sends you a reprint request, add a personal note with thanks and comments about their research or any other linking topic. Those few seconds of extra time spent are much better than just dropping the reprints in an envelope. Always answer another's correspondence, even if it is only with a brief note and an apology for not being able to answer more fully. Courtesy counts.

To aid in corresponding, I keep a notepad handy in which I write down those often fleeting thoughts or questions that I would like to pass on to someone else. I then organize these notes together by who is to be the recipient and refer to them as I write a letter or e-mail. I try to do this when the number of my notes gets to a certain number, which tells me it is time to block off an hour or two to write. I, thus, do not forget or neglect keeping in some contact and the contact often has some scientific context and value or point of interest for the recipients.

One important tool in corresponding and in building and maintaining your network is to put together an address book. This can be the older styled paper copy or an electronic one. The latter can be added to and edited quickly and neatly and can be copied for safekeeping. I use both, having started a paper copy before electronic ones were convenient. These contain the name, street address, phone and fax numbers, and e-mail addresses of people I correspond with. It also has grown to include a lot of people who are resources that I keep more intermittent contact with. An address book is very handy because you can easily refer to it to call or write to someone with an idea or question. It is not old fashioned or a dated thing since networking is not out of style if you want to succeed.

Giving out and receiving business cards should become habitual. They should always be handy; kept in the briefcase, wallet, or handbag so that they can be given at any opportunity. Offering your business card should be part of any introduction. Do not feel reticent about asking for those of others. These two actions should be part of every meeting, especially including the informal ones during research conferences. The exchange of business cards is very important in certain cultures. Offering yours is seen as a sign of respect and friendliness. In collecting those of others, you should build a card file or other handy way of keeping them in some organized fashion. The information on those of important new contacts should also be transferred to your address book.

This information can and often should be shared among your network, but you must do that with care. Your network is your resource and it can be beneficial to all in extending their own networks. You, however, should not do this haphazardly or cavalierly, as the trust and friendliness that the members of your network show to you may not be mutual among themselves. Tarnishing a relationship can happen if two in your network do not get along or if some other negative interaction occurs. If you have any doubts in passing on contact information, first check with that person to see if he or she is comfortable with that.

Another aspect of communication in building networks relies on the fact that the interesting scientific talk at a conference is not limited only to the presentations. The in-the-hallway or coffee-break discussions also contain much science. Often it is what is going on in labs at the moment, not months earlier when presentations were submitted to meeting organizers. In these casual discussions, it is easy to meet and get to know other chemists. At the same time, they learn more about you. Talk of hobbies, shared interests, and personal information in the scientific context is no different than in other aspects of life. These human connections naturally predispose others to look upon you more or less favorably, depending on what others think of you. Generally, the reactions are much more favorable.

I once had a conversation with someone at a conference because he was sitting and doing the large, difficult crossword puzzle that the New York Times prints daily. I sometimes did those puzzles myself, so I commented on my enjoyment in also doing them. This puzzle worker turned out to be a Nobel laureate. Even though we never collaborated, I think that this is a good example of how an individual can increase his or her chances of networking. Every interaction with another chemist (or scientist of any discipline) is part of building your network.

One attitude in interacting with others in this way is the give-and-take, the reciprocity of it. An individual best builds a network in which the others perceive the interactions to be two-way. Each person receives and gives information, ideas, and opinions. Selfish people who only look to gain from others will soon find that their networks are limited and not dynamic.

Networking often involves exchanges of more than ideas and communication or samples and standard compounds or reprints and copies of articles. The use of resources and the time to utilize them are also exchanged. If you can do things for people because you have resources that they do not, then you can offer them and strengthen your network.

Examples I have done of this sort include running molecular modeling programs, collecting fluorescence spectra, determining molecular weights by size-exclusion chromatography, performing literature searches and others. I have received a variety of things in return and not necessarily from the same people. Your network can be an open exchange of things given and received. Things do not need to be given gratis, but the bartering must be positive and not niggling over equity in the trading. In the long run you will gain as much from others overall as you give to a point where the balance does not matter. You and those in your network all gain.

The converse results can happen and have an effect on your research if you receive greatly more than you are willing to give. Your reputation can be a calling card that tells others that you are not a good potential member of a network. You are not helpful or collaborative. Your interests are only towards your own gain. This is not only a situation that you do not want to be in if you desire the best chance for success, it is also more difficult to avoid than its opposite. People remember much more intensely what they perceive as a slight or a wrong done to them than they do the positive alternatives. One mistake that you do to someone else can negate several good things you have done. There are no real tallies kept, but people do have a sense of the balance if it is too skewed away from them.

I am reminded of what was my first exposure to a gathering of eminent scientists. While in graduate school, I and several of my research group attended a large international meeting. One day as another student and I wandered at the exhibition that was part of the conference, we ran into our research advisor. Being near noontime, he asked if we had plans for lunch. We said no, expecting the usual foray to an eatery. He led us through the conference center to a banquet room. It was the annual awards recipient luncheon. As a past winner, our research advisor was invited and could bring a guest (or two). We were seated among many eminent scientists, those names on books and seminal papers that we had read. Before our advisor introduced us, he mentioned one name and said "Do not talk of your research with him. He is known to steal ideas and go back to his laboratory and scoop you." The name was that of a very well-known researcher, the discoverer of many important things in our field. We followed his admonition and also became aware of how unethical behaviors can lessen research opportunities. In a later conversation with our advisor, we learned that this particular scientist was a brilliant lone wolf who could thrive on his own. This is a rare type of success and the alternatives in collaboration are much more likely.

Building a network takes time and the timing must also affect one's attitudes in interacting with others. No one when first met is ineligible for future membership in one's network. You must remember that the graduate and postdoctoral students of today will in a few years be professors and scientists scattered throughout academia, industry, and government laboratories. Neither should one focus only on trying to network with eminent names, the easily recognized leaders in the field. These brilliant scientists should not be ignored, but times do change. A little-known professor today may build a strong reputation. Also distinguished scientists eventually retire from research and may rapidly lose some of their value in a network. So networking must be both dynamic in that it is always being done and is unlimited in who is involved. It never hurts to have a strong link to someone before they become a Nobel laureate or other award winner!

Another thing that helps to build a network is to rarely turn down any request to be involved in the workings of a journal or conference. I use a rule of reciprocity as guidance. The workings of science involve both giving and receiving. If I take advantage of a journal to publish a paper or of a conference to present one, then I should also be willing to serve as a reviewer or chair a session if I am asked to do so. My papers are reviewed by others and I benefit from having published papers. If someone else wishes to also benefit by having a published paper, then I should be able to review theirs. These sorts of tasks also have the added advantage of sometimes letting me see new developments before others since reviewers get the first peek at research results.

When you must turn down a request, take the time to let the person who offered it know politely why. Everyone can accept reasons such as other obligations that coincide with a conference or too tight of a schedule or other already-promised review articles being written. Courtesy matters in building better contacts. Although you might have to turn down one opportunity, this leaves the door open for other later ones.

Many of these ideas are variations on what people do to build friendships in their non-technical lives. These are only modified and adapted slightly to fit within the context of the scientific world. This is no coincidence! Once again it must be emphasized that scientists are people and feel the same emotions as others. Getting people to like (or dislike) you is much the same in science as in other areas.

In building and maintaining your network, remember to do the human touches that you use in your daily contact with family, neighbors, and friends. Your professional network is made up of people, many of whom will become friends. Listen to them in conversations about personal interests. Remember that these facts are as important as the analogous learning about professional aspects of the friends in your personal life. They are the flip sides of similar coins.

Just as you know that your neighbors are an accountant, a school teacher, and a carpenter, you can learn that this environmental chemist teaches scuba diving as well as the martial art of aikido or that chromatographer likes to read science fiction or that the theoretical chemist likes to cook gourmet dishes. In both parts of your life you learn of family and other things of importance to each individual.

3.3   Collaborative Research

Over the years I have been fortunately involved in a wide variety of collaborations. In this context I refer to collaboration as it is found in the United States, Canada, Japan, and a few other countries, in which the work is often informally structured. Even if it is formally structured through shared funding, the structure is limited to two, three, or only a small number of institutions and individuals. The large formal structures found in Europe, the consortia, will be discussed separately in a section at the end of this chapter. These can operate internally along the lines espoused here for informal structured collaborations, but the funding, legal, and bureaucratic defining points can force differences.

The types of research and the diversity of those that I have worked with have led to me being asked more than once "How did that happen?" The specific answers are as numerous as the number of collaborations. The general answers, however, have common points and themes to them. Several of the more important ones occur in the starting stages, while a few others occur within the collaboration itself.

Although I will discuss collaboration in the context of a full partnership in a research project, there are also lesser degrees of working together. These reflect the lesser degree of doing some of the things discussed below. Do not limit yourself to only major sharing of your ideas and talents. Helping others with an insightful idea or information may not result in a collaboration and coauthorship on a publication, but it is rewarding in helping colleagues and to foster better science. An acknowledgement is not a bad thing. This cooperative attitude, if done often and with enough people, will result in many fruitful collaborations.

If you do enough of this, some will become collaborations. Think of yourself as casting out a variety of seeds. Some will germinate and the variety will give you many results. The more you share, the better the chances for good collaborations and an interesting variety of them, too. This is part of the taking of initiative, the networking, the diversifying, and the continuous learning described in separate chapters. These all coalesce into collaborative thinking.

Setting Up a Collaboration

The first suggestion is to keep your eyes and your ears open to opportunities. Every publication that you have read, every presentation that you have heard, and even every conversation or correspondence that you have had, is a potential collaboration. If your scientific interests have no boundaries, then you can potentially contribute to any other research project. You only need knowledge, insight, and creativity, among a few other things (this is said with a smile, as those do not necessarily come easily).

One of the most common points in creating a collaboration is taking the initiative to start an interaction. Volunteering ideas or observations, hard-to-find standard compounds, or ideal samples to prove a new technique have all been the impetus of several studies and of the beginnings of collaborations in my own work. A collaboration may be a partnership but it also may not happen if the others involved do not get enthused about working on it. A good idea may be enough to offer, but also giving resources that make it more likely to be successful and enjoyable will increase the chance.

Too often someone reading a paper or listening to a presentation may think "I would also have done this as a further step" or "I wonder why they did not do that?" Many of these readers/listeners then pass over these thoughts. They assume that the researchers involved are working on those continuations or that they have already been explored. If the paper is from one of the "big name" researchers in the field, the reader/listener may think it is presumptuous to make contact solely to pass on speculative ideas. These are not always good assumptions. Do not just let these thoughts become fleeting mental exercises. Make a note of them, and then follow up at a more convenient time.

Very often a leader in a research area has not thought of the idea you propose or could not do it because of limited resources which you think are not a limit because they are not for your own work. I had a collaboration with Dr. Maximillian Zander in which perhydrocoronene was used as the spectral matrix. It is an inert and flat-shaped molecule. It can, therefore, trap other flat molecules like the polycyclic aromatic hydrocarbons between its layers as it crystallizes after melting a mixture of it and the "guest" molecule. This started when I asked Dr. Zander why he had done only a couple of compounds in this fashion in work two decades earlier. His reply was "I ran out of perhydrocoronene." That is a simple and reasonable answer for anyone but me. I had a supply of several hundred grams of it from some other research work. This might have been the world's supply of this compound, but it was mine to share. I sent Dr. Zander several dozen grams of the perhydrocoronene to use in further matrix-isolation spectral experiments. Three new papers in this type of matrix-isolation spectroscopy soon resulted.

If the areas of research that the reader has thought of are indeed being pursued, it has been my experience that the investigator, when contacted, will then readily say so. This may be accompanied by a request that that information be kept in confidence until a paper has been submitted. If not the areas of the research ideas are not being pursued, a lively discussion of the idea often ensues. This can lead to collaboration.

At a minimum such an approach usually leads to an exchange of fruitful ideas and a tacit delineation of whose research work will explore which areas. So if you suggest an idea for joint research, it may be novel and if not and the person already is working on it, then the only consequence is that you must keep quiet about that on-going research until it is completed.

Some people might baulk at this offering of good ideas so freely. My response is that if you cannot perform those experiments yourself, then why not help the others explore? If you can perform the research, then why not collaborate with someone who does similar work so that you can divide the tasks, get the work done sooner, and report your brainstorm and its results?

As scientists, we all have a keen sense of discovery and learning, building a better understanding. This element was a common theme when talking with other researchers about their collaborations. If all of the collaborators share the same goals of investigation and discovery, then success is more likely.

One aspect of this which I have found is the enthusiasm from someone when you give them new areas to explore, other variables to look at. If you suggest an idea for research, almost always there will be acknowledgements in any subsequent papers. If the researcher is generous or your ideas were significant to making the experiments succeed, a coauthorship often occurs.

Another key element of being limited by a boundary in thinking is that the only interesting work must be in the comfortable areas where one's past experience has been. A person working on HPLC may only think of new areas in HPLC research, even if these are as diverse as retention mechanisms, selectivity of a separation, and new bonded phases or packing materials. An atomic spectroscopist may only think of working in sample digestion, atomic adsorption, or plasma emission.

This attitude limits. Many of the real innovations, the breakthroughs, are when expertise in two complimentary areas comes together. Douglas Lane, an atmospheric environmental chemist gives an example from his research of his atmospheric chamber that could simulate environmental conditions very well, but did not have coating on the sample collecting denuders to work optimally. At a conference he heard a talk by Lara Gundel describing her denuders that worked much better. Her experiments needed a better atmospheric chamber. Putting the two together resulted in a very fruitful collaboration of many years length.

Being Open to Offers of Collaboration

If someone develops a new technique or instrument or possesses unique chemicals or samples, then others will often want to collaborate on other ideas after word has spread through the first publication or presentation about it. There is a chance of much good science happening through these contacts. The recipient must first be open to the suggestions and not give any offer quick and cursory "No" as the reply. This may be easy because the new idea disrupts one's own plans and diverts the focus away from one's own ideas.

Many highly interesting collaborations happen when the exchanges of ideas cross disciplinary boundaries. When a technique is first applied to another discipline, rapid discoveries occur. These can even be ground breaking. A whole field of astrophysics opened up when infrared, UV, and fluorescence spectrometers were coupled to optical telescopes. Today hundreds of astronomers and astrophysicists scan the heavens finding new evidence of planetary and interstellar molecules.

Another common boundary in thought is that geography is an impediment. If the other researcher is far away, collaboration seems difficult or impossible. Face-to-face or hands-on collaborations seem much more natural. That may have been true years ago. With today's global communications and shipping resources, researchers from every part of the globe can work together. I have had several collaborations on both research projects and on writing review articles in which my other colleagues and I never met. This is unfortunate from a human interaction aspect, but still doable as far as the science.

One of the largest examples of this wider thinking and willingness to cross boundaries was the discovery of the macroscopic route to produce fullerenes. The researchers involved were astrophysicists looking for ways to simulate interstellar space. They sought to produce infrared and UV spectra similar to features seen in interstellar space. These are ascribed to be due to vast amounts of very large polycyclic aromatic hydrocarbons. These researchers vaporized graphite in hopes of creating carbonaceous material in this ethereal environment. They produced a soot which, instead, contained a brownish-red soluble material. This new form of carbon was not what they expected and did not move them closer to their goal. The rest of the story is the spawning of thousands of research projects centered on C_{60} and the other fullerenes, carbon nanotubes, endohedral structures, and the many other new carbon structures found in that soot. All of this science arose because a few astrophysicists played with carbon and looked at it slightly differently.

This cross-disciplinary openness also encompasses the mirror image of having initiative to approach others, which is being receptive when others approach you with their ideas. Ideas and proposals from other researchers should be met with interest and curiosity. Someone in another field may see opportunities that you may not. This leads to discussions and sometimes does require some study and learning of the other discipline so that you choose collaborations wisely. The amounts of time, standards, or samples you might have are all limited and cannot be given to every scientist with a proposal. A wise selecting, however, can lead to discoveries in astrophysics, oceanography, environmental science, and many other diverse fields, as my own research has shown.

Part of this receptiveness is also teaching the scientists in those other areas the fundamentals and important details of your own area. Doing this is time consuming and takes efforts, but if everyone involved in a collaboration does this then each both teaches and learns the expertise each member has. This leads to a partnership and often an on-going series of experiments.

Another area that is extremely limiting to collaborating, but which can often be either subtly, or is even worse if blatantly done, is to limit one's interactions only with scientists who are from a similar scientific, cultural, or social background. Much has been said and written about the need for diversity in the sciences, so I will not touch further on this.

There are many ways to surmount this barrier and also bring personal benefits to everyone involved. Not limiting any interactions or collaborations on any criteria except technical ones is key. Collaborations are learning experiences and are dynamic for everyone. Each member learns and teaches the others in areas of weaknesses and strengths of knowledge. It can be difficult to avoid the unconscious inclinations to work with similar personalities. That is human nature, but it must be remembered that scientific knowledge and creativity are not limited to any one type of person.

When personalities match and there is open communication, a true collaboration results that is often long term. The exchange of critical ideas and speculation on results and future experiments happen. With comfort with each other, there is no need for guarded ideas.

135

A final point on open communications is that to delineate roles, expectation, and schedules in the beginning stage of a collaboration always help. Each collaborator brings individual talents and knowledge that must be meshed into the framework of a working team. In order to keep the collaboration on-going and vibrant, issues such as who will write manuscripts, the ordering of authors, and who else is acknowledged must be discussed and agreed to early on. If not, hurt sensibilities can arise that will kill any future work. It is amazing how scientists with dozens of published papers can still be offended or hurt by being the second or third author when they thought they deserved another slot. Even the overlooking of an expected acknowledgement can cause friction that limits future collaboration. What many authors think of as a minor politeness can stifle interactions if hurt feelings are a result.

It should be apparent from all of these points that personalities and attitudes are keys to successful collaborations. Interacting with others is much easier and can often be very positive in its dynamics if the personalities are compatible. Note that I use compatible rather than similar. Many types of personalities can work well together, even if they might not be close enough to build strong friendships. There should be mutual respect and a willingness to both give and receive ideas.

Several scientists I spoke with recounted bad collaborations. These often were one-sided exchanges where resources, materials, rare chemicals, and ideas were given and little was offered by the other collaborator. One organic chemist spoke of a giving of ideas without the feeling of a returned respect for them. This soured the relationship somewhat. Others spoke of collaborations where their expertise and resources were used to launch on-going research programs which started as collaborations. After the initial phase, however, the recipients did all of the work on their own with little or no acknowledgement to the collaborative start, leaving their former collaborator feeling that his research had been pirated as they proceeded on without him.

These areas can be very touchy ones for individuals. A sense of teamwork and collaboration must include communications at all stages to ensure that there are no negative feelings. Scientists, as many parts of this book emphasize, are still people. If a person thinks that acknowledgement of ideas or other gains should be made and they are not, it is a mismatch of thinking. The onus should be on the recipient collaborator. In a really dynamic partnership, this happens when all of those involved in the project are really involved in all aspects with little domination by any individuals.

Are collaborations necessary if they are more difficult to deal with than one's working alone? An extremely creative and adaptive thinker might not think so. But collaborations often lead to many unplanned and unforeseen research opportunities. They can focus the talents and resources of several researchers on the work, giving faster and better results. They also create a network of colleagues that are resources for one's individual research and as conduits bringing opportunities to be involved in societies, journals, and conferences.

Consortium – A Large Formal Structured Collaboration, European Style

In the United States, academics are encouraged to do more applied and useful research now than before. Industry concurrently has a very short research timeframe. This leads to a balance and synergy. Industry does not fund much research directly, but some is done through trade groups, such as the Electrical Power Research Institute (EPRI), the Asphalt Institute, the American Petroleum Institute, the American Chemistry Council, etc. They do, however, supply many ideas for long-term research that their laboratories will not do. Academics in the know use industrial chemists for sources of such ideas. The academics' funding is theirs and does not require any other participants.

After writing an equivalent essay on the topic of collaboration, I received numerous correspondences from European scientists describing how different collaboration is in the current unity of Europe. These described a surprising situation, as I had collaborated with scientists informally from the UK, France, Germany, Sweden, Poland, and Hungary through the years. The consortium is now a defined and required structure for much research funding in Europe.

137

3.4   Diversity in Science – Being Open-Minded

It is natural for each of us to develop friendships and other personal interactions based on our own personalities. Our criteria are biased towards those characteristics that we see and esteem in ourselves. This can be good because compatible attitudes and ways of thinking lead to more open interactions. These, in turn when involving other scientists, lead to more creative and wide-ranging exchanges of ideas. When we are comfortable with someone else and less conscious of potential negative reactions, then there is more possibility of thinking in new ways and looking at an idea or problem differently.

It may be that a person naturally reacts positively to those like him or her, but the apparent corollary that someone different should generate a negative response should not be true. The thinking should rather be that a similar personality elicits a very strong positive response and another may be less so, but still positive, at least until interactions determine how compatible the two personalities are. This is especially true in an intellectual undertaking like scientific research. We cannot tell how capable or insightful a scientist is by appearances.

I, however, know at first hand that this happens. I have a vision problem and many people react to this by assuming I am less intelligent and less capable because I falter at things done easily by someone with normal vision. An organic chemistry professor told me that she often gets the same response because she is a young and attractive woman. We must ask ourselves why this is so? The outside does not show what sorts of brains we have or how we use them.

There is also a need not to be exclusive in one's interaction to only those most similar to oneself. Being limited only to those similar to us, however, can be a limit in our potential as 3cientists. New research often involves looking at things differently. A variety of colleagues can give rise to new work. The people familiar to a field may all think similarly, while someone with a different view may ask "Why must we do it this way? Why not try to do it that way?" Broadening one's group of colleagues and collaborators to include a wider variety of people also broadens the perspectives and ways of thinking that are available. It is a way to combat "tunnel vision", thinking in which everyone thinks alike and cannot be open to alternatives. Science needs alternatives to move forward.

One trend in chemistry favors this. The proportion of women receiving degrees in chemistry grows every year at all levels. According to Chemical and Engineering News (April 7, 2003, pg. 44) in the year 2002, for instance, in the United States more than half of the bachelor's degrees were earned by women, 55 percent. Women earned a few fractions of a percentage higher than this of the master's degrees and 47 percent of the doctorates. With these numbers, it will be more and more difficult for male chemists to not include women in their networks, collaborations, and other interactions. This pushes both a social and psychological change in chemistry, as Myers- Briggs personality typing shows differences between the sexes. In my opinion, this can only be good in broadening the perspectives and possible directions in research.

Linda Osbourne, an environmental chemist with Heritage Environmental, stated both her own experiences and the benefits of a diverse working team "In regard to diversity and acceptance, I have been fortunate enough to be surrounded by supportive people. It is true that men and women are very different in their approach to tasks and in how those tasks are completed, but they can often compliment each other. I have a strong conviction that a balance of talents and the right chemistry between a group, regardless of their diversity, is the key to a successful operation along with a good work ethic. Within this organization, from the top to the bottom, I feel that I have been supported over the past 18 years in all of my endeavors. My female status as a research chemist has not really hampered my career. Alternatively, it has balanced out; sometimes it appears to be advantageous and other times a detriment with no overall net effect. Passion for your work can overcome many obstacles and that appears to apply to the multitude of diversified faces. My advice for career development is to find something that ignites you and go with that! Nothing else will matter."

This emphasis on a passion for the work, liking what you are doing was echoed by Jocelyn Hellou, a marine environmental chemist with the Department of Fisheries in Canada. "Go after what you want, persevere, and work hard." This attitude is similar to that of many scientists.

Another aspect of diversifying and accepting others of both genders and of all nationalities and ethnicities is to involve them in research, in professional societies, in producing journals, and the other parts of a dynamic research field. Too often in the past, different categories of people were essentially excluded from the core of a research field. At different times, women or those from lesser developed nations or of certain educational backgrounds or of certain ethnicities or for other categorizations will be treated differently. This has narrowed the scope of the science that was done and thus limited the number of talented people involved in moving the science forward.

Scientists have some advantages over other fields. Much of scientific research communications is written. The books and journals carry only a name. This often tells little of the person. Gender can even be unhinted at if initials are all that are given of an author's name. We can develop our images of each other solely on scientific capabilities if we initially become familiar through this fashion. How often do we develop a mental image of someone from their research and find that that mental image is correct? The answer is very seldom if ever. We develop our images of the scientist regardless of the person.

If we accept the quality of scientific work that is read in papers and book chapters, then there should be a similar esteem when meeting that same person or hearing her or him giving a presentation. It should not be in the vein of comments like "It is disconcerting and even irksome to think that some in the audience are not thinking of my research, but rather looking at my legs or breasts" as one woman scientist put it or "I hope never to feel that anyone sees my skin color before they hear my words." as another put it when referring to ethnicity.

Fortunately in the past two decades this limiting has lessened for women in chemistry, although it is nowhere near non-existent. There are programs in many national chemical societies that aim at reducing the biases and making the profession more inclusive. What advantages are there to doing this? What can an individual scientist do?

There are several advantages for a scientist to be diverse in the types of people that make up her or his network or group of colleagues. One of the most obvious is the greater range in ideas and opinions that can be received. This is especially important given the wider-ranging number of research fields, the greater complexity in each field, and the rapid paces of discovery in each field. A more diverse network means an increased chance of becoming aware of developments in other fields that may be of use in one's own research.

Another advantage is more altruistic or can be described as better for one's field of study. The number of highly talented researchers in one's field will be greater if there are no barriers to involvement by anyone. It is very backward for scientists to assume that women or those of certain ethnicities or nationalities cannot be excellent researchers. This was the attitude in most sciences until recent years. Science a hundred years ago was almost exclusively done by men from Europe, North America, and a few other nations such as Australia, New Zealand, and South Africa. Today's science cannot afford to exclude so much talent; yet psychological and cultural barriers still exist. Every researcher has the potential to continue the change from a localized science to a globalized science.

3.5   Using a Mentor

When one is starting one's career, a lot of challenges lie ahead. Even though they may seem daunting, there is one resource which is convenient and which can help make the struggle easier. This is by finding a mentor to show the way and to tell of how she or he overcame similar problems. How does the young scientist find a mentor? What qualities should be looked for in one?

There are two types of mentors that the young scientist may need, technical and non-technical. The first group would be those more experienced scientists who can give advice and knowledge on the science. Experiences in writing research proposals, in teaching, or in attracting graduate students generally would also fall within this category since the issues are not related to the individual's personality and attitudes. A technical mentor is relatively easier to find as scientific skills are the predominant criterion. The second type of mentor is those who can give advice on career decisions, personality clashes, ideas on networking, and all of the other non-scientific aspects in a career.

As an aside, many of the same issues in looking for a mentor are also those needed in looking for a graduate school research advisor. A research advisor is a specific type of mentor and teacher. The difference is that the research advisor serves as the research science advisor in a formal role. This cannot be uncoupled from the personalities of the two individuals that are involved. Compatibility of a graduate student and research advisor are key to success in learning/teaching and collaborating. It is the most important aspect for a student in graduate school. Other issues that are large include the paramount one of what type of science is being done in the professor's research group and the lesser one of what funding and resources are available. For other researchers outside of that context, choosing technical mentors is much less absolute and can rely solely on scientific knowledge.

143

Technical brilliance is often one attribute that younger scientists recognize and naturally think of as a mentoring characteristic. This may sometimes be the case, but it also sometimes may not be. The personality that favors mentoring is not tied into the qualities that lead to technical brilliance. A willingness to listen and give advice can come from anyone. The wisdom of understanding experiences, leaning from them, and being able to pass these on are some of those of a mentor. Observing people, understanding their thinking and motivations, and then coming up with possible actions that deal with these and help proceed are others. Little of these talents are inherent to doing good technical work. Choosing a mentor on scientific brilliance may lead you to relying on a cold, self-centered, opinionated person who gives bad advice.

In order to understand why this may be so and to accept that idea, I would like to turn to the analogy of sports where the star athlete is often a rather poor coach because she or he often did things readily. Such a person understands less of the fundamentals needed to perform. Often it just seemed to happen for them. Although they whet and hone their skills, the natural abilities they possess did not give them the experience of learning how to do their sport. Their practice and skill building moved them to the higher levels of performance.

The more marginal athlete, one who may have reached the higher levels of competition through hours after hours of practice, is much more aware of the skills needed and what types of work will build them. Many successful coaches were not star athletes, but minor ones. Those like this in science may make excellent mentors because they had to study, work, and think harder to get where they are. A brilliant scientist, on the other hand, may remember everything ever read or heard on a subject and easily be able to assimilate and to coalesce these into new, creative ideas. The ideas from a brilliant scientist, therefore, may not be useful because the work in doing research was simple and easy. The things that that person thinks need doing may be woefully skimpy for another because their talents differ.

A person seeking mentoring needs to look more for personality traits and attitudes that will set up a comfortable dialogue. This lets each open up, with questions and with answers. One important trait is listening. Another is open-mindedness, being able to look at several alternatives rather than readily picking one and deciding that that is the only correct one. If one wants good advice because she or he does not understand the various options, then relying on someone who only looks at one or two ideas and decides quickly may be setting up for failure. If the decisions are important and complicated, then a mentor who can think thoroughly about the situation and options is really what is needed.

Although one would like one's mentor to be similar in personality, attitudes, and background because that is most comfortable, it is not always the best course. Mentors are chosen to help and to ensure wise decisions. A person with a different personality, with different attitudes, or with a different background may be more objective. It brings in different perspectives and many other options. Choosing someone similar may not yield unbiased opinions. That person's experiences may cause the advice to be too limiting because the solutions and options offered will also be limited.

In some cases, however, a mentor must be either someone like oneself or someone who can greatly relate. For example, a woman should seek out another woman as a mentor when dealing with gender-related issues in the workplace or on how to balance work and personal responsibilities and time. Experience with the issue does matter.

One does not need to pick only one mentor. One may, and most likely should, have an ever-changing herd of mentors throughout one's career. Having ideas from several people may make the choosing of options easier. The weighing of the various alternatives can be done with more examples and more experiences in dealing with similar decisions. These are people who can be relied upon for advice and suggested directions in a variety of topics. At any given time, one or two may be selected

144

due to her or his particular experience in certain things. A difference in perspectives can give you more information and possibilities to choose from. Or perhaps, a mentor who is good at career direction may not be able to help balance professional and private lives.

Relying on a mentor also requires you to have several personal attitudes. You must be comfortable with seeking help. You must be willing to listen to the advice even when the suggestions are difficult. You must follow through on the advice and apply it.

Mentors are volunteers, who are asked to be such for your benefit. As such, they deserve both respect and thanks. The respect comes in both taking their ideas seriously even when they run counter to what one thinks and to not react with any negative emotions. This may be especially hard if the mentor has given a negative message. When you are told "You made a mistake." Or "You are not putting enough effort into the work." It is natural to bristle a little. The mentor, however, is giving guidance to help. Sometimes this just must be negative. Setting up the trust and dialogue with mentors that allow this is needed. There can be times when every person needs to hear such messages.

There are a few things that make it easier to set up a trusting relationship with a mentor. Talking with her or him often on a variety of topics helps. These may be minor issues as well as major ones. If the work environment seems uncomfortable to either of you, set up the meeting in a more relaxed atmosphere. If at work, do it over coffee in the morning. Having lunch, breakfast, or dinner are options. The first seems most natural, but the last may allow for more time. If you drop in on your mentor impromptu, always ask if she or he has time to talk. If not, set up a mutually convenient time. Tell your mentor briefly what you need to discuss so that they can gather some thoughts in the interim.

Remember when hearing this kind of message from a mentor that it was difficult for her or him to deliver. Remind yourself that delivering a negative message does not gain anything for the giver. This should remind you that the only one who will benefit is you. Then move onto asking your mentor why they think that this message was needed. Listen carefully to the explanation and if you need to, ask for rephrasing and clarifying until you fully understand it. Then discuss possible changes and solutions. This leads on to a path of accepting that a drastic change is needed and that you can start making it.

Follow-through and follow-up are two bits of advice to do after receiving suggestions from a mentor. Follow through by choosing the suggestions that seem most valid. When the results of this occur, let your mentor know and thank her or him again for the help.

3.6 Being a Mentor

BEING A *Mentor*

In academia the training of younger scientists is an inherent core part of the system. The purpose of a graduate school program is to perform high levels of research while utilizing graduate students. This helps the students learn how to become beginning scientists, carrying on their own careers. This is the first phase of education for a scientist and ends upon receipt of the masters or doctorate degree.

The next phase is a much longer one, whether in academia or not. This is the "learning to be an eminent scientist" phase. The purpose of many of these chapters is to help in the areas that need to be done in order to succeed in this phase. Are there other resources for the young scientist seeking guidance in getting through this phase? After replying with a resounding Yes! I must say what the most obvious and common one is. This is looking for mentors among one's more experienced colleagues, those who have gone this route before you. That in itself is the topic of another chapter – choosing your mentors. For this one the other side will be presented, that is "How to be a mentor and why it is rewarding".

The word mentor as used here describes someone who gives advice and wisdom. It is meant to also include the formal role of being an academic advisor for undergraduate students and a research advisor for graduate students. In the first role, the professor does give guidance on course work that will build up the knowledge of the students so that they can continue onto graduate student work in the research areas that they choose. The academic advisor may also assist in the application for and selection of graduate schools. The relationship is formal, but not rigidly so. The students are not required to adhere to the advice.

The graduate research advisor's role is very structured and entails many responsibilities. Much of this advice is mandatory and the relationship inherently has much less give and take. The student is not only in a subordinate situation, but the advisor is the sole boss and often controls the monies that the student receives. These things give the research advisor a much different role than just being a mentor.

As far as the how-to-be part in being a mentor, it can be very easy for most scientists. Giving opinions and advice are almost natural for the person who thinks analytically. Looking at facts, assessing them, coming up with explanations, and coming up with solutions and plans for the future are all parts of the work of being a research scientist. These all involve coming up with opinions and ideas and often giving them to others to implement. If we do this as part of our work, then doing it for the benefit of others in their careers is not that different. It involves the human aspects instead of science, but as I said before, each of us has gone the route that awaits the young scientist.

So the answer to the question "Is being a mentor difficult?" I would answer that it is very surprisingly easy. It happens normally just because we each gain experience and knowledge during our careers and we also are normally asked for our opinions as part of being researchers. Being a mentor is only combining these two things. This is true and happens in all three venues of research chemistry. In academia, the more senior professors often advise younger ones on how to succeed. So the mentoring is not only professor to student, but also professor to professor. In industry and government laboratories, too, the more experienced scientists help the lesser experienced ones.

I think for many people this also happens almost unconsciously as they answer the questions of others about how they dealt with certain things. In the early years of a career these queries will mainly be on smaller issues, such as dealing with procurement of a new instrument or how to get approval for a conference trip or what to say in certain types of presentations in order to be most effective or dozens of other specific topics. Answering these types of questions is easy, but also gets the individual into the frame of mind of accepting to be a resource. Generally without even noticing it or trying to, a person becomes used to giving advice. The questions become more complex and even more personal as one continues to give advice, but the growth is slow enough that no undue pressures are raised. If he or she has had a lot of experience and gives good advice, then the role becomes that of a mentor.

As a researcher grows in experience, others will seek his or her opinion on other issues. This is particularly true if the "senior" person has had direct experience with the topics that are being asked about or if she or he has an open image. That is an image in which others perceive that talking with someone is not unwelcome, will be kept in confidence, will be listened to, and will yield ideas and perspectives that help in the situation. Each of these is important in themselves.

Most people think of the last one as being of paramount and almost of sole importance. The others, however, set up of real communication and dialogue as well as a comfort level is needed on the "junior" person's part to ask for guidance. Asking for guidance is not always an easy matter, especially if the issue is a personal one, a difficult one, or a major one. Thoughts of changing jobs – to another organization within a company, into another career area or profession, or to another company – are ones that many people will only discuss in strict confidence. A listening recipient makes the other more comfortable by showing real interest and concern.

This communication and dialogue must be open. The mentor must not feel obliged to only pass on wisdom and ideas for the future. Sometimes there must be advice given about the present. Sometimes there must be advice given that is unbidden. The person seeking the advice should have made it clear in the initial stages that there will be trust, acceptance, and listening even when the message may be negative. There are situations in which the mentor must point out shortcomings or ways in which things were not done well. When this is needed, the mentor should feel comfortable in doing so. It is difficult to tell someone these things and even more so if resistance, rejection, or an angry outburst is possible. The mentor must be tactful and also offer positive suggestions and possible solutions, but the negative message should be firm.

Voluntary mentors find this most difficult. Research professors do as well with their graduate and postdoctoral students, but have the advantage of possessing the formal responsibility of being an advisor. This mitigates things to some degree, but even then tact, careful wording, a positive attitude, and giving alternatives must still be done. For the voluntary mentor in any of the three venues, these are needed as well. This situation does not fall only on the mentor to deal with. As I mentioned earlier, the younger scientist must have trust and faith in the mentor. I describe that more in the being mentored chapter.

In giving advice, a mentor is asked to do so from her or his experience. This means both the immediate personal experiences and those that have been observed for others. A mentor should give more than one view, more than one perspective, and more than one option. If the aim is to help the other person, then passing on as much as possible lets them choose through the wisdom and guidance of the mentor. The mentor becomes a guide, not a leader. The other person learns what to think about and how to reach a decision.

Is mentoring time consuming? "Very often not" is the answer. Many of these discussions take only minutes and can be done in bits and pieces over a period of time. More in-depth conversations can readily be done both in the workplace and outside of it. Time spent talking over breakfast or lunch or dinner can be very good for mentoring. For those with whom it works better, having coffee or drinks after work also is one way of having the time and venue for a longer mentoring conversation.

What are the rewards in being a mentor? There are several that I can readily list. First, it is always satisfying to feel that others value your opinions and knowledge. Being thought of as wise and a source of guidance reflects on one's own successes and the good points of personality. Being regarded by others as a good source of information and opinion is an acknowledgement of one's career successes and knowledge. It is also gratifying after the fact when one who has received the advice returns with news of the success and thanks for the useful help. This sense of altruism and ego-gratification is enough reward for many scientists.

Another reward is the feeling of satisfaction at returning something back. Each of us has been mentored and guided by others in our own careers. A way of acknowledging those that helped us is to help others in a similar fashion. There is an inherent continuum in our profession of mentoring that goes back for decades and decades. Each of us is the end link in a chain that may even go back for centuries of science as a profession. So in addition to the altruism of helping another who is "where I was years ago", there is also the added sense of being part of this continuing chain.

Seeing other scientists take one's advice and use it in progressing in their careers is also a reward. This is inherently built in for a graduate school advisor who proudly acknowledges that certain scientists are former students. The same sense of pride can infuse anyone who mentors other scientists and sees them succeed later.

Finally, from a more personal perspective mentoring makes one think of good ideas and reflect on one's own career. All too often we take ourselves for granted, either through humility or familiarity. In describing her or his own experiences, a person can often realize what good things were done over the course of the career. Most people focus on the most recent accomplishments or a few past high-water achievements, seldom remembering the bulk of the good work done.

This may at first seem to be an odd situation since it would seem that an individual must know best her or his own accomplishments, but human nature creates selectivity. Try listing your own dozen best accomplishments. If you are like most people, it will be heavily weighed towards the most recent years. Then think of what you did for earlier years. A lot of good work has typically faded from one's own memory. Mentoring often brings out similar results of recollecting a lot of less-remembered accomplishments that resurface while one passed on advice.

In summary, mentoring is fairly easy, is not a major burden on either time or effort, and has both personal rewards and benefits to our profession.

The Academic Difference – Projecting Mentorship and Attracting Students

The volunteer mentor is the norm in government laboratories and in industry. In academia, the mentor role is more formal. I have touched some on this and the overall aspects are similar.

There are, however, a few important differences.

One is that in being a more formal role, there are responsibilities for the mentor besides giving advice. The research advisor must ensure progress towards the degree. There also must not be an over-extension of the student's time – a longer than needed period in which the experienced student's talents are milked for another paper or two. A perception of this can both embitter those students of yours who feel that they are being taken advantage of and deter potential students from joining your research group.

Another large difference that professors must deal with is that they must be mentors. It is a role that they cannot avoid. They cannot succeed without students. Not only is a dynamic research program less likely, but their universities are educational institutions. A young professor who does not attract graduate students is not likely to attain tenure. Even those with tenure will receive pressure from their department chair and deans to gain students.

So part of being a research advisor is attracting students and preferably the better ones. What attracts the better students? Besides the obvious and primary reason of interesting and vigorous research, the reasons depend more on personality and attitudes. These may not be very observable by the beginning students who are choosing research advisors, but the more senior students know. They readily pass on information to the younger ones and the younger ones will listen because they have so few sources of information.

Being pleasant and kind is not a prerequisite to attract students. Some professors may be known to be very demanding of their students; pressing them to their utmost. Some of the most loyal students and former students had advisors of this sort. What they mentioned, though, was that the demands were never unwarranted and unfair. The advisor assessed the potential of the student continually and expected that full potential to be gained and applied to the research. The students realize that this approach pushed them to do their best and were thankful that the advisor had made them do it.

If you gain a reputation for working with your students to make them attain their degrees and the skills to start a research career, in whatever personal style matches your personality, then students will be drawn to you. Conversely, if your reputation is that you think of the students only as a cheap pair-of-hands to perform your experiments or that you do draw students' time out to gain more from them, then students will avoid you.

3.7   Behaviors

3.7.1   Personalities and Styles in Dealing with Others

In this chapter I hope to highlight the importance of personality and attitudes in good research work. This will then lead into more specific chapters that delve into how these can affect research and how they can be altered. These have been touched on in earlier chapters, particularly in collaborating and networking, since personality plays an important role in both.

Contrary to the viewpoints of many research scientists, science involves as much psychology and sociology as any endeavor that involves interactions and communications between people. Many collaborations and rivalries, many feuds and friendships, and many examples of generosity and penury are found in the history of scientific research.

Many scientists think of themselves as dispassionate seekers after the true working of Nature. Much of the nonscientific population also believes in this image, sometimes to an even greater degree. This logical, coldly emotionless scientist is stereotypical and appears in much of literature and in many motion pictures and television programs.

In reality, scientists are as varied as people are in general. They have the full spectrum of personalities and attitudes. Are the varied characteristics prevalent in successful scientists reflected in their success? Yes and no, are both the answer. Yes, in that scientific success is not limited to any one type of personality or even general types of behavior and attitude. A successful scientist may be anywhere in the ranges of being penurious or generous, friendly and outgoing or shy and reserved or standoffish and cold, soft-spoken or full of bravado, more comfortable working alone or more comfortable working in a team, close-minded or receptive, etc. These traits reflect the individuals as people who happen to have chosen science as careers.

The answer "no" must also be given because certain personality characteristics can increase or decrease the chances of success. This is the same situation as it is in many professions, as well as in life in general. An outgoing person finds it easier to speak and so giving presentations comes more naturally. She or he networks more readily and so knows more of what is currently being done. An open-minded scientist finds it much easier to have productive discussions with others about research. A narrow-minded individual, particularly if this is accompanied by a judgmental nature, will soon find colleagues and peers being very close mouthed about their work. The chances for networking and having robust discussions about the direction of this individual's research are thus less likely.

What traits aid in success or what traits hinder it? Science by its nature requires some almost paradoxical combinations of characteristics. There are opposites that delicately balance to highlight both the individual and the collective aspects that are needed. For example, the individual or small groups of individuals working together are touted for discoveries and gain the accolades such as are epitomized by awards. With the increased complexity of many technical areas, the ability to be a valued individual contributor on a large project team has become much less possible. No one person or small group of people has the full resources and experience for projects such as collecting and assessing the data from an interplanetary spacecraft or mapping the human genome. Working together as an integral part of these large teams requires individuals to aim for less egocentric goals.

One of the balances that a scientist must achieve is in the personal satisfaction in the acquiring of individual credit without this being so overly strong that the work seems to be mainly that individual's with a surrounding cast of many supporting characters of lesser importance. If a researcher is seen to be too focused on gaining credit for his or her own self, then others react to this egocentricity. These others become less open in discussions of current research and less collaborative and cooperative in doing the parts of the project that they must do.

This dilemma is the same as it is in many aspects of life and in most other professions. A touch of humility blended with self-confidence is a hard mixture to achieve.

I was once told by an eminent scientist "None of us knows everything, even in our own specialties". I think of this often. It reminds me that no one in any field is so dominant in knowledge or skills so as to be intimidating and overawing of others. This also infers the corollary idea that anyone may have a useful contribution to make. On teams, each person has a role that relies on and supports the work of the others. Not denigrating others is as important as not overly touting one's self.

Of course there are individual traits that are not negative and so do not need to be moderated and which must be present for success. The motivation to learn is one and the drive to explore new areas is another. The sciences are not easily learned and understood. They are not stagnant. New discoveries must be kept up with. New theories must be understood. New instrumentation and methodologies must be acquired and used. Hard work must be done and diligence maintained in order to keep current.

Some of the important traits needed to maintain knowledge are an acknowledgement that keeping current is important, that being organized in one's time allows literature reading, and that the literature must be archived in such a way as to be a ready resource when needed. Some of these attitudes and the approaches that accompany them are described in the various other chapters.

An individual's attitudes determine how they think and approach work. Most scientists have a strong streak of optimism in them. This coincides with the thinking that new research can discover and can solve problems. Sometimes a pessimistic attitude may arise relating to a specific issue, but overall the forward-looking attitude that comes with doing research overrides those down-moments.

Experimental scientists are not so optimistic that they always wear those proverbial rose-colored glasses. They are realists, too. Optimism is based on a drive to succeed and the expertise that is built up to make the research succeed. Several researchers described that dealing with the failures and difficulties was a challenge at times, but that the moods were transitory. One highlighted the book by Norman Vincent Peale "The Power of Positive Thinking" as a great resource and inspiration. I will not delve into the topic much further than to also say that many scientists are religious and cite that as a strength that helps them.

Contrasting with these individualistic traits are collective efforts such as the exchanges of ideas while work is being planned or done and the open exchange of preprints. The first of these is done with the often-unstated understanding that there will be acknowledgement of any contributions in any subsequent publications. The second has an inherent understanding that preprinted work is confidential since it is not yet accepted for publication or available to others. Neither of these understandings is usually expressly agreed to, but an honors system is kept by those doing these things.

There is a caveat, as expressed by Richard Mathies of the University of California at Berkeley, "Everyone is screwed by somebody at sometime." What he means is that in the course of an active career there will be a time when each scientist trusts another and is then harmed by that. It may be a one-sided "collaboration" where one person gives more in time, resources, ideas, and efforts and the other takes the majority of credit. It may be unethical behaviors, such as the stealing of an idea to gain a claim of discovery. It may be being treated with a lack of respect because of one's gender, heritage, lack of seniority, or other criteria. It may be due to any number of reasons.

A person must trust in order to network, collaborate, and to be an active member of the scientific community. This leads to the chance of a less-than-scrupulous person to behave badly. The lesson learned is to retain some small amount of wariness to see if the trust is met or not. If it is not, then end the interaction as rapidly as possible. Learn from mistakes and move onward. Do not get obsessive or ascribe all such things to a prejudice. Bad people are not necessarily biased. They often only find convenient targets in whoever may be used to advance themselves.

The breaching of these understandings by using another's ideas without acknowledgement or by prematurely spreading word of another's unpublished work without consent are both strong taboos in our scientific culture. Word of mouth will soon spread from those scientists injured by these breaches to others working in that field. The scientific reputations of those breaking such confidences are not often harmed. Opportunities, however, are lost of hearing of the latest hot news before it is formally published or of engaging in as much collaboration as might have been possible.

If one is involved in the free exchange of ideas and information, however, the opposite results often happen. Word spreads that this individual is a good source for insights, review of plans, different perspectives, and other valuable ideas. It is highly valued that these are available while retaining confidentiality. Although it is not assured that good things happen from this, but it has been my experience that it often does. The scientists who involve themselves in these sorts of discussions also are the first when nominations for the chairing of symposia sessions, for journal editorial boards, for society committees, and for other roles are needed.

These aspects of the dynamics of science also rely on certain communications skills that are rooted in personality. Listening for information without being judgmental or without always wanting to introduce one's own ideas before another has finished speaking requires both listening and patience. It is often difficult not to lose attention of another person expressing ideas without starting to think of one's own response, but speaking too soon cuts short on collaborative dialogue. This can be detrimental among peers, but is devastating to a constructive discussion when one of the individuals is a student (or another scientist of less experience) and the other is a very experienced scientist.

If the negative traits are foremost in this situation, such as not listening fully and espousing one's views over any others, dialogue will be stifled. Although this can happen among peers and colleagues, it is less likely to be detrimental unless the person is overbearing or denigrating in expressing his or her views. In the situation where there are great differences in experience, however, this can silence and intimidate the less-experienced person. This can lead to resentments. These interpersonal issues are often key factors in whether someone stays or moves into another research position. A research institution or university chemistry department may lose many talented people due to the lack of personal skills and attitudes of one individual.

This overbearing treatment can be a limiting factor in an academic's career, as word soon spreads among graduate students that this person would not be a good professor to work with. The brighter, more capable students can often be selective in choosing a research professor, thus the lone-wolf-style of professor loses out in the competition for the better students.

Thus, having the wrong attitudes and interactions can harm the career of the individual, the careers of those working around him or her, and even the functioning of the places where they work. In contrast, good attitudes, collaboration, and an emphasis on working positively help the careers of all and the successes of those places.

Along with adhering to the right attitudes and avoiding those wrong ones goes the need to be open minded with others. As is highlighted in the chapter on diversity, people who do not think like you do may still be a benefit. They look at problems with different perspectives. They assess issues with different degrees of importance. Seeking diverse opinions is good; avoiding them leads to "tunnel vision" which may limit your options. This may lead to bad decisions or ones that do not cover all the important aspects.

Once again I refer to the Myers-Briggs personality types. Your one of the sixteen lends itself to being comfortable with others of that type. That does not mean there is less value in others of the remaining fifteen. They are only different. The title of one of the books on the subject sums it up well; "I'm Not Crazy, I'm Just Not You".

3.7.2   Dealing with Those on the Dark Side – Difficult and Worse People

In the course of a career, it is almost impossible to avoid dealing with unpleasant situations and people. This is true in any of the three venues, although the likely situations and the types of personal interactions in each will differ. In academia, a difficult department chair or dean can be the issue or there can be politics within a department. Analogous situations can happen in the government laboratories with the added pressures and negativity of funding woes.

The last factor is very common in industry, as are organizational disruptions due to mergers, company restructuring, changing business orientations, and the downsizing or closure of facilities. In all of these situations, tensions often run high. People are less happy and more prone to negative behaviors. Competitiveness, for an example, increases in the times when funding worries become an issue as some people act in what they think is self-preservation. There is a scramble for funding and credit that precludes thinking of the welfare of others.

In some instances these situations pass. If this is the case, your aim must be surviving through the difficult times. This may involve a focus more on your work to divert your attention from the stresses. Alternatively, it may entail becoming more involved in activities outside of the workplace.

In other cases the negative things are on-going and chronic. In this second group, the attitudes you take must be either to learn how to deal with them and persist in that position or to decide if that situation is worth staying in. There is a balance between the two that must be weighed. This is, on the one hand, between those parts of the work that still bring satisfaction versus the chances of a new situation being better. Added to either side are the intricacies in switching positions either within your organization or company or by changing to other ones, and the other factors involved in changing positions.

It is an unfortunate aspect of human psychology that negative events and people generally have much more impact on our emotions and memories than do the corresponding positive things. This makes dealing with them much more important. If 99 % of your career is positive or neutral, that 1 % of negatives can dominate your outlook and your perspective on how well your career is going. Even after years of a successful career, a few months of a terrible situation involving bad colleagues, bad team members, a bad supervisor, and a bad manager can change the course of a career from where it was to one where any possible alternative seems wonderful.

160

I will describe some of these negative people and situations in this chapter. Additionally, there are a few that are unique to teams and those are described in that chapter. I use convenient names that are of my own choosing or are ones others have used in discussing negative people. These may seem trivial to some, but the characteristics are real types of bad behaviors and the suggestions for dealing with them are based on the experiences of real research scientists, including my own. These examples are not all encompassing for bad personalities and behaviors. They are intended to give a pattern that shows insights into the motivations, attitudes, and thinking that gives rise to these bad interactions.

I have focused on personalities that affect the working interactions; those of the supervisor, customer, colleague, and subordinate. There are other personalities that are negative in themselves. Bigoted and biased people are one example. Their views on gender, race, ethnicity, religious affiliation, and other aspects are highly skewed against some people and for others. This behavior is unethical and in most countries it is also illegal. As pointed out in the chapter on ethics, any unethical person is detrimental. Anyone who falsifies data or steals money or diverts it towards personal use is of a negative type and should be avoided totally. These types of people can do reprehensible things in any role and in any situation.

Ignoring their misdeeds will not be a long-term solution, even for those many people who want to be non-confrontational, and who choose to act as if it is not happening, and of their concern. If the person is still in your working environment, at some point there is a chance that you will be obliged to work with them and thus leave yourself open to becoming tainted if their actions are discovered. It may not be prudent to directly confront such people or even to make managers aware of them. That depends on the individual and circumstances. But being scrupulous oneself and trying to keep a personal record to avoid incrimination are minimal efforts. If one has a confidential mentor, discussing the situation may lead to a solution and provides corroboration that there was concern over it. If there is sufficient trust, a manager can be made aware.

Negative situations can generally be dealt with in some fashion to make them better. Some, however, cannot.

These might be due to organizational structures and culture which are not very manageable by an individual. These preclude accomplishing much because the situation or attitudes cause many people to behave badly. Trying to change an individual's attitudes might be possible in general. If, however, they behave badly as a consequence of the situation or because the situation allows and supports bad behaviors, then working for change is most often futile.

When dealing with negative people or negative situations, be choosy. Do not fight every battle that appears. Many frustrations are not going to change whether you are anxious, angry, frustrated, or sad. For example, although a company may claim a tough financial situation making it necessary to reduce the number of employees or have little or no increases in salaries, those same financial circumstances seldom have as much effect on the number of managers or their compensation. In most cases, if the executives of the company are reduced or suffer real diminished compensation, then expect to see hens with teeth or a blue moon. The executives and managers make the operating rules and that means they protect their own interests first. This may not be fair to you and the other workers who are vulnerable and less compensated, but that is the situation.

Being upset over such things may seem to be only human, but when there are other stresses, they have to be ignored as much as possible. Adding weight to your own mental load does not help you deal with other things that you might be able to effect. Save your energy for those battles where you can either induce some change or in which there is enough personal satisfaction to make the effort worthwhile. Spending your time and energy on a losing battle can be worth it if standing true to your values is important.

You cannot change certain things, such as many company or institutional policies or government regulations. Trying to do so often just uses up a lot of energy and time without gaining any changes. A person who fights every battle will most often either be Don Quixote fighting against windmills in meaningless efforts or like a knight fighting a huge fire-breathing dragon in one-sided, almost suicidal conflicts. Strategic retreats save you from a lot of distress, headaches, and heartburn.

Many issues are not worth the contentiousness or cannot be won. Decide if the issues are really important enough to risk a strongly negative interaction or confrontation. A current saying says "Do not sweat the small stuff".

Also trying to fight a battle under someone else's rules will only get you trounced. Assess the situation. If the person is in a position of great strength, you cannot win. For example, if the bad person is a highly recognized lone wolf, you cannot do much since the technical brilliance the person displays is an accepted part of the organizational culture. Showing that this person is not a good team player, is not collaborative, or even that he or she steals ideas and credit will have little credence. In that last case, it is difficult to prove, is readily ascribed to jealousy by managers, and means little to managers because they really care little about who made the key discoveries as long as the work gets done and the inventor was someone in their organization.

Unless you have an evil side, you cannot outdo someone else who does unscrupulous or unethical things. Trying to get revenge may seem like a thing to try, but if anyone treats you unfairly, doing something similar to them will most often backfire. This gives them ammunition against you and may even give the impression to others that the person was justified. If someone unfairly takes more credit than is due, you cannot make things right by trumpeting your own or others' real efforts. This makes you look like the egocentric person. You must try to deal with negative behaviors by ignoring them if possible, minimizing their effect, or counterbalancing them with your own positive efforts. It may not seem correct or fair, but in the end you do not succumb to doing things that you detest in others.

By coming in contact with so many individuals, it is only natural that some interactions are not positive. Although we would wish every other person to be thoughtful, helpful, collaborative, sharing, considerate, and several dozen other good attributes, this is not the reality in science as it is not the reality in any other aspects of life. Some people are negative inherently, being rude or selfish or pessimistic or insensitive or bad tempered or any other negative characteristic. Sometimes it is easier to interact with them if you understand their motivations.

Since interacting with negative people is inevitable,

the question must be asked "How can one accomplish what must be done with the least amount of detrimental impact?" This all depends on the specific individuals, on the tasks to be done, and on the situation in which the interaction occurs. Given that there are so many scenarios, it is only possible to give general ideas, advice, and a few examples in order to set an attitude in which one must try to have in order to deal as best as possible with a negative relationship. Some of the categories described can be found in other roles than those listed. The underlying attitudes are the same, in general, so that you can deal with that type of person in whatever role they are in. There are books dealing with this topic that are listed in the bibliography. Here, I will try to give examples more tuned to a research chemist's situation.

The different interactions are divided by the relative roles of you and the person with whom you are interacting negatively. In some, the person is in authority, a more powerful position. This may either be someone in a supervisory role or someone for whom you are doing work. In other situations, it is as peers. In others, it is a negative interaction with someone in which you are in the more powerful role.

Bad Supervisors

There are a variety of bad supervisors, as there will be bad personalities for any of the potential roles in one's career. Each must be dealt with in its own fashion, with some particular care to the individual personality involved. In some of these types, acquiescing to a good degree may seem unpalatable, but antagonizing a vindictive supervisor means the end of any career chances with that organization.

The *short-timer* is a supervisor who has been in the role for a long time, is bored with it, and who only waits for retirement. This is the short time, the years until their career is over. This type of person seldom thinks anything is important. Relying on or expecting actions or support from such a person is both futile and frustrating. Looking to others to help and to fill the void is a solution. This is particularly so in the case of a cross-organizational project or team situation. Other supervisors have a vested interest in success, too.

The *climber* is a supervisor who always is looking towards the ladder upward in the organization. His interests in you are mainly limited in how your work may make him look better as a supervisor. This is not necessarily bad for you if you are aware of it. A climber will highlight your work if it brings the spotlight on himself, too. On the bad side, a climber will not share blames nor assume any responsibility for work other than the stellar successes. If there is a situation where blame must be given, a climber will either stand by to avoid any or even worse, he or she will deflect any blame onto whoever is a convenient recipient.

One subset of climbers is political supervisors. They want to keep everyone happy. "They shake every hand and kiss every baby" is how one researcher put it. Their behaviors to please all customers and all other managers can be confusing and inconsistent to the point of being paradoxical. One story was recounted in which the scientist had a truly phenomenal year, highly successful with lots of pleased customers and impact. There was one lone exception. A customer who ranted and raved over poor work from the scientist. This amounted to less than 1 % of the scientist's time for the year. His supervisor, in order to please everyone, gave the scientist a poor annual performance review that included a salary increase that was much more than double that of anyone else. The supervisor could tell that one dissatisfied customer, however, that there had been a poor performance in the review.

With a climber, you must be expected to share some credit and assume more blame. This is unfair, but it is the reality with someone who needs the spotlight. Fortunately, climbers do not often stay in the same position within an organization for very long. At worst, they move up the organizational ladder. More often they transfer into other areas.

The *Peter Principal* is a supervisor who is described by the Peter Principle, that in a large organization people rise until they reach a level in which they cannot competently perform. This can lead to a deadwood supervisor or manager. This is someone who has been promoted to that Peter Principle limit.

A *weathervane* supervisor is inconsistent. His or her attitudes change with whatever is going on. Job duties and requirements change as often as anyone else in authority talks to the weathervane. This can lead to an overburdening if things are not defined. Having roles, responsibilities, and schedules written down and agreed to helps. Thus, each iteration is only a modification due to changing circumstances. This makes the job very varied, but if you know that will be the case, then you can deal with each new day. The main chore is to ensure that you do not end up having to do everything that this kind of supervisor has agreed to. That may require much more time and resources than is possible.

Bad Customers

When one is working in the situation where the work is at the request of another, the bad customer can arise. This is most common in industry, but can also occur in other venues. In academia, it can be a professor or research advisor or fellow graduate student. What is meant by the bad customer? There are several common types. One is the person who always waits until the last moment before requesting help, creating a crisis situation. This type of person can be handled by frequent reminders of the time it takes to perform the work needed or by having the overall workload being coordinated and prioritized by involving both the requestor's and one's own managers or supervisors.

An ancillary bad customer to this type is the customer who always deems her or his work as top priority and in a crisis. This is also sometimes manageable through coordination and prioritization by involving managers. This customer must be constantly reminded of progress being made and the realistic timelines that are involved. They also need more communication in the front end so that they do not delay in starting the work and create unnecessary crises.

Another bad type is the customer who cannot or simply does not define the criteria of scheduling or the tasks needed to be done, thus leaving the work open to continual redefinition, rework, and continual complaint. Up-front planning, including definition of timeframes and specific tasks to be accomplished must always be done. These can be redefined, mutually, as the project progresses, thus there are still defined goals.

The Know-it-all Customer

In certain situations, the customer may think that he or she can define the exact work that needs to be done. This assumes technical decision-making over the work without assuming technical responsibility. For example, if the customer asks for an infrared spectrum rather than asking for a solution to a particular problem or telling the synthetic chemist to try a Wittig reaction to give a certain desired new molecule. A person should attempt to induce a dialogue on the technical issues with the aim of assessing the reasons why those specific tasks were chosen. In doing so, the fundamental aims of the work will be exposed and defined. If these match the set of tasks chosen, then the work is valid. If they do not, then suggestions of alternatives can be given along with reasons why they are more appropriate.

Analytical chemistry is often looked upon by chemists of other disciplines as the "poor cousin". It is the discipline that everyone else can do as well themselves without the need for an analytical chemist. This view is particularly strong among physical and organic chemists who think separations and spectroscopy are easy and simple because they have done them. This often leads to the requestor of the work not leaving any room for the person doing the work to vary the plan or add things that may be an improvement in quality, lower cost, or decreased time spent.

Although I use analytical chemistry as an example from my own experience, other fields suffer from the same lack of respect. If there is only a shallow familiarity with the work, a customer can assume they know all that is needed to make the decisions. Enlightening these people with the details can help. Try to touch on any of the more intricate or complex issues and how they are important in the context of the customer's goals.

The Moving-Target Customer

Vague goals, schedules, and priorities mark this type of bad customer. Projects are ill-defined. Definition of a project's scope, duration, timing, priority, and the expected results are all things that should be discussed at the project's start. Continued review of these and redefinition as needed should be done as the work continues. Any redefinitions should be mutually agreed on so that all people have the same goals and expectations.

The Perpetually Unhappy Customer

This is the customer who wants more data, more accuracy, or more precision. Often this is at no additional cost and with little or no allowance for more time. What this type of customer needs most is a reality check. This is done by describing in great detail, with schedules, timelines, cost breakdowns as needed, the efforts needed to do the main work itself and any additional efforts needed for the extras. Reiterating these details until there is realization can be frustrating, but most people will eventually accept that you cannot run ten sequential chromatographic separations of one-hour each in less than ten hours. Sample preparation and data evaluation will add to this.

Bad Colleagues

There are numerous versions of bad colleagues. I will touch on a few to give a flavor of them, to help understand their negative behaviors, and to help deal with them.

The self-seeking colleague, whose wishes are to collaborate only for his own gain, can be difficult to deal with because playing a game for credit is a part of how he or she does things. Once credit is given, it is hard to totally undo unless plagiarism or idea stealing is blatant and provable without any doubts. The best defense includes keeping meticulous records of ideas and experiments; copying key supervisors, team leaders, and managers on all progress reports; and in general making everyone aware of who did what and when.

In any team effort, there is the potential for shared successes. There is, however, also the potential for one team member to garner more credit than is due. This can be divisive and cause any future teams involving this person to be doomed to failure.

The Unreliable Teammate

In the team situation, tasks are both shared and disbursed. The responsibility for accomplishing all of these tasks must also be shared, since ultimately the team's success or failure relies on everyone contributing what they must do. For many people this can be difficult because few people want to seem overbearing and intrusive. In delegation of tasks, however, there should be interim reviews to see if things are going as planned. This both deals with the less-motivated or less-diligent team members and also with those situations where there may be a problem that is holding someone up on their tasks. In the team situation this can more readily be done as there is both sharing of responsibility for a project and sharing of many tasks. Suggestions to aid in helping the person be more reliable and more involved include helping on specific tasks or setting targets for each individual's set of tasks.

Incompetent or technically limited colleagues also fall into the category of unreliable people. The tasks they are given cannot be done properly without help. Obviously when work involves such people, then that support must be part of the plan.

The sharing of roles in this team situation can vary according to the need. In an extreme case, cross-training of another team member can create a redundancy that ensures that someone can and will do the tasks. This job sharing can be implemented in a very positive fashion if the reason for the recalcitrant team member's not meeting goals is due to too many other tasks on other projects.

The Arrogant, Inflexible Colleague

A good team values each member's capabilities and accomplishments. Each person contributes to the overall success. There is no problem with a diversity of contributions because each person is contributing their own utmost. In order for this to happen, an altruism and sense of equality must be held by everyone.

Insecurities and resentments result when any member starts to dominate the team in an egocentric fashion. One example is an arrogant individual who thinks he or she knows more things and better than the others what the needs are. This close-mindedness combined with an attempt at overawing any input from others leads the team into failure. The members often only participate minimally until a goal is reached, never doing their best efforts at work that they do not feel ownership of.

Personality clashes can be very intense when team members finally revolt against an arrogant colleague. The pent-up frustrations lead to personal attacks and arguments.

The Bad Subordinate

This is a very tricky situation that is not as easily handled as may seem at first glance. Most people think that being the boss, supervising someone, gives them power and authority that can be used to overcome any difficulties. This is only partially true and the use of that force may lead to a worse situation. A subordinate who is forced to "do it your way" can submit grudgingly. This leads to resentments that can take many forms. Some are varieties of sabotage; some are grumblings and the spreading of rumors; some are subterfuges where it only appears that this person is following your guidelines.

In today's employment situation, legal issues or those involving a union can be very difficult for a supervisor. Even academicians must now deal with unionized graduate students. Treating a graduate student badly can no longer be done with impunity.

Reference
Peter L.J. (1969) The Peter Principle, William Morrow and Co.

3.7.3   Ethics – The Right Things To Do

Ethical conduct is something we increasingly hear of in the news with reference to science. It is used in reports on "discoveries" based on fraudulent data, on the utilization of genetic engineering, on potential repercussions of chemicals in the environment and to human health, and other areas.

What is meant by the word ethics? In general usage, ethics is defined in the dictionary on my desk, the Webster's New Collegiate Dictionary, as "the discipline that deals with what is good and bad and with moral duty and obligation". In science this translates into the good and bad behaviors that other scientists expected you to have. These ethics are demanded by the scientific community in order for it to be viable and dynamic. In this chapter I will touch on those general issues. For specific areas, such as the ethics involved in the publishing process or as a supervisor, there are further discussions in those chapters.

In the past several years there have been a few very well-publicized examples of fraudulent scientific research. In some cases this has been by deliberately falsifying data to report a desired result. In other examples, the interpretation of results has been biased upon an aim to end up with a particularly desired result. Neither type of behavior is acceptable.

Two of the most publicized examples of research based on fraudulent data were submitted by groups of scientists, the claims of superconductance in organic molecules and the synthesis of several new artificial elements. In both cases, later investigation implicated that the principle investigator committed the fraud. The other workers on the projects had little real involvement or oversight. They abrogated their responsibilities as colleagues and coauthors, allowing the fraud to go unnoticed. This is taking a passive role in being a collaborator and not verifying the results. This is especially heinous when the research is groundbreaking because the results cannot be accepted only by glancing through a manuscript. Every author listed on a manuscript must remember that his or her being named means something. Any future accolades or denigration are to be shared.

In a reflection of the growing need for more inclusion of ethical behavior in research, the American Chemical Society has recently formed a committee on ethics (Chemical and Engineering News, April 21, 2003, p. 40). Its responsibilities include reminding society members of the expected ethics through mails and articles in the Society's publications and to provide education on the topic. It joins a variety of ethics-oriented groups that already operate within the ACS that deal with employment, hiring, proprietary rights, and other issues.

The journals published by the American Chemical Society now contain a guideline on ethical behavior. It touches on many aspects of research and the ethical standards that are expected by editors, authors, and reviewers. Its definition and origins are given. "An essential feature of a profession is an acceptance by its members of a code that outlines desirable behavior and specifies obligations of members to each other and the public. Such a code derives from a desire to maximize perceived benefits to society and to the profession as a whole and to limit actions that might serve the narrow self-interests of individuals." It goes on saying that the sharing of knowledge to advance science must be done at the sacrifice of some personal gains.

There are several specific obligations described for scientists in the three roles, researchers submitting manuscripts and other scientists acting as reviewers and editors. The main concepts are in confidentiality and impartiality. The publication process must be a closed process with those involved maintaining secrecy of the manuscripts until they are published. Any possible conflicts of interests or biases, positive or negative, must be recognized an$ dealt with appropriately.

Discussions with colleagues about research prior to its publication are confidential. The free exchange of ideas can lead to some very innovative thinking. As an example, an experimenter might perform some novel experiments. The fruitful results might not be explainable. These could be gained by discussions before publication. Some explanation reached in this way could support the new experimental results that the experimenter might not have thought of.

These exchanges, however, are not in the public domain. The information is not published yet. The manuscript could become caught up in a long, drawn-out review process or rejected. Anyone who was made privy to it and passed that information to others would risk the due credit to the original authors of the discovery. Unknowingly, this can happen. These other researchers may refer to the first work as "unpublished results" in their manuscripts, but theirs becomes the first in print. The recognition for the initial published work becomes theirs, not the real discoverers. Other later researchers forget the circumstances or are not aware of them and cite the first published report. Credit for the discovery then is attached to the authors of the first published report and not to those who first did the research.

Editors and reviewers are included in this sort of

confidentiality. They are made aware of this new research as part of the publication process. They bear the responsibility of keeping the work private until it is accepted for publication. Even then the ACS guidelines state that only the title and authors should be published and only as in the context of future articles to appear in the journal. All other referring to the work and its details should only be done with the authors' express permission. Many journals require written confirmation of this prior to allowing mention of such private knowledge. That this is unpublished information or in press with the list of authors should also be noted in any manuscripts by an editor, a reviewer, or a fellow scientist with this preprint or personally discussed knowledge.

Conflicts of interest or biases must not be hidden, especially in the reviewing of manuscripts or research grants. In a very active field, this can be difficult. A reviewer may know the authors and their work. They may be considered friends or competitors. In either case, a reviewer should try to set aside that personal knowledge in assessing the manuscript. If that cannot be done then the offer to review the manuscript should be denied and it should be returned quickly to the editor.

An even-handed appraisal of the research described in a manuscript is even more difficult at times. Reviewers are chosen for their expertise in an area. They often know personally everyone else doing the higher quality work in that area. One must be on-guard for biases for or against the others based on personality, friendship, enmity, and other personal aspects. The review process relies on this integrity and there should never be any compromising of it because of personal issues. Poor research should not be published even if it is done by a friend. In fact, it is somewhat of a disservice to do that. An even better approach would be to honestly review the manuscript and help the friend achieve the good work needed through detailed comments. Contrarily, good work should not be rejected because of enmity.

A more subtle form of a conflict of interest is a bias against a manuscript because of the origin of the work. The nation, institution, and other affiliation of the authors should not be an issue that weighs negatively or positively on the treatment of the manuscript. This origin might be reflected in a less-than-acceptable level of technology, reagent purity, equipment, and other technical issues or in the quality of the writing and grammar. These aspects can be criteria for acceptance or rejection.

Due credit must be given for significant contributions to a project. The magnitude of the contribution determines if this is a coauthorship or an acknowledgement. Conversely, coauthorships should not be taken lightly so as to include any person who is even slightly involved in the work. I must admit to be overly generous at times with coauthorships. But it seems to me that if the decision is debatable, then it is better to err on the inclusive side. With that, the author listing may grow longer, but the chances of bruised egos are less. A rule of thumb reiterated by several scientists is "think of yourself as the other person. How would I feel if I were or were not acknowledged? If I were or were not cited as a coauthor?

Quoting from another research paper is legitimate if it is cited. If it is not, it is plagiarism. The introduction sections of manuscripts or review articles sometimes contain bits of this. It seems that authors think the writing someone else does is good and that they cannot do better. I once reviewed a manuscript which had a very characteristic opening paragraph. In describing the occurrence of the larger polycyclic aromatic hydrocarbons "are reported from literally the bottoms of the oceans to the dust clouds of interstellar space'. It contained certain phrases and words that sounded very familiar. It was copied from one of my papers, but without citation. If I had not been one of the reviewers this writing would have been usurped by the authors of the later manuscript.

The ACS guidelines mention two examples in the ethical listing of coauthors. Colleagues who die before the publishing of a manuscript should be noted as deceased with the date of their passing. An additional notice about coauthors is that no spurious or fictional authors should be listed. The fact that this is specifically mentioned reflects that some authors do not seriously consider coauthorship.

In the less formal world within an organization within industry or government, the same issues arise. If there is a collaboration or team effort, shared credit must always be given. Even when someone does work for you, such as an analysis or data evaluation, you must cite and credit them for that work. You did not do it.

174

3.8   Teams, Teamwork, and Leadership

Many of the attitudes and attributes that are described in various chapters come together in one part of a research chemist's work. When someone is involved in work as part of a team, they need to be reliable, to be diverse, to communicate, to plan and create the numerous aspects of these topics. Although the individual chapters touch on some of the things involved in working effectively on a team, in this chapter I hope to bring all of those ideas together. I also hope to emphasize the extra aspects that are unique for each topic when applied to a team.

The team in this context is a *project team*, as opposed to an organizational team. The needs of a team within a structured organization are covered in the chapter on supervising and managing. A project team is a transient group that works together to solve some scientific or technical problem. They are often diverse in work functions, parent organizations, and backgrounds. Additionally there will be a focus on the leading of a team.

This combining and blending of so many things makes the effort more than that of any number of individuals doing research work each on their own or even more than that in a collaboration. There is both a synergy of these things within each of the individuals and also between the individuals among themselves. A highly productive team often is not just good researchers working together. There is interplay and teamwork that make the efficiency and effectiveness greater.

A team, by its nature, should be a group capable of doing more together. If not, then why work in a team rather than as individuals? Coordination of efforts to increase efficiency, collaboration, and creativity among the participants which brings diversity of thought and new perspectives, and a sharing of knowledge and resources are some of the things that happen on a good team that help the group exceed what the individuals could do if only working alone.

As a temporary working group, the affiliations and loyalties to the parent organizations of each member may be an issue. This can be a problem both originating in the team members and in the managers in their organizations if the project team has not been given autonomy and authority to act in its own best interests. This arises especially when the project team relies on funding from the various organizations directly rather than through its own approved budget. When going into a project, one must understand who the customer for the work is and whether the boss that can be relied upon for decisions and definition of the direction and goal.

Trust – Working in Unity

A team is made up of individuals. There are many factors that can make the members diverse. In today's global scientific and work situation, this diversity can get increasingly large as the team gets more disperse. In the local situation, everyone works at the same place and for the same organization. These individuals may differ in temperament, attitudes, motivations, and job function. Conversely, they will speak the same language, relate to each other's personal lives because of the similarities of locale, and probably be from the same culture. At the other end of the range, projects may involve people around the globe. These may be different in those aspects that are the same for a local team.

Getting everyone to understand their other team members is important because any similarities or understanding builds a sense of commonality. This empathy leads to trust in working together. Any dissimilarities can lead to mistrust and potential conflict if they are left undefined and not shared. It is important to build the understanding and accepting of these differences as only the variations between people and not as things that impede communicating and working together.

Preston Smith and Emily Blanck, two experts in team building and interactions, write in great detail on the issues that influence the workings of a global team. Some issues that are not of importance locally can be major issues. Time zone and geographical differences can make direct communication difficult. They point out that someone in the Northern Hemisphere will not consciously accept as a year-end deadline approaches, that someone in Australia may be in the midst of a summer holiday (as is true for those in New Zealand, South Africa, Argentina, and other austral nations). The European or North American mindset is that summer is in June, July, and August, and work is planned around a holiday in those months. Many North Americans do not account for the summer holiday periods which are the universal norm in nations like France, Spain, Italy, and Germany.

Another corollary example, in Europe and North America, the work week is Monday through Friday. Deadlines are often set for the end of a week. Team members working in a Moslem nation will gear towards a work week ending on Thursday, as the holy day is Friday. Such deadlines put an extra pressure on them because of the one fewer day. This is on top of the several hours later in the day for them when plans are formulated and schedules are set. They can feel that they receive tasks later and that these are due earlier than for the other team members.

These issues must be dealt with in the initial planning stage of a project before they arise as problems. Once perceived as a problem they can be divisive and lead to a less effective team. A set of operating principles must be created and agreed by everyone. Members should be aware of the time-zonal differences. They should know the regular work schedule which especially includes the sharing of working times of those with less regular hours, either through shift work or those such as the four ten-hour-day ones that are becoming more popular in the United States. Out of office messages and message and telephone call forwarding must be habitual options when possible. Far-flung teams are easily disconnected and lose their effectiveness.

Smith and Blanck emphasize that face-to-face meeting should be arranged to happen in the planning stage. Every participant gets to know the others, gets whatever individual issues and concerns dealt with, and the group creates the plans, schedules, and targets. A first meeting later in the project is nowhere near as effective. If a face-to-face meeting is not possible, then a very high-quality teleconference is an alternative. The visual connection is important. A phone conference call or e-mailing to get a project started does not build the trust and unity needed.

Motivation – A Common Goal

When individuals are interviewed about any highly successful teams they have worked in, one theme is reiterated. Having a common goal that strongly motivates everyone is a key. If you are the designated team leader or in a position where assuming the leadership seems natural if there is no designated leader, then you must figure out how to not only involve everyone but to pull them into being highly motivated.

As discussed throughout this book, the emotional makeup of people varies. Therefore, their motivations vary, what drives them towards success will vary, and how they behave in a team will also vary. A leader must try to find what gets the best out of each team member. This takes work and extra efforts, but that is part of leadership. A real leader does not try and mold everyone into sameness. This de-emphasizes the individual talents, skills, and knowledge. In research those are valuable. Research relies on creativity. If you treat everyone similarly, and in an even worse case if you try to make everyone fit into a follower's role, you will destroy creativity. The result would be that all innovation would have to spring from you.

As the team's leader you must often paradoxically not be seen as the team's leader. If there is too strong of an image that you are the leader, then people may be less motivated or may rely too heavily on you rather than solving the problems that arise. This reliance can reduce the initiative and creativity that the team members should be using and it can greatly slow down progress. If people think they need your approval on everything, then you become a bottleneck that defines the pace of each step.

Most people mistake leadership for dictatorship. It is easy to become authoritative and take control if you are designated to be a project team leader. That would also in most likelihood reduce the team's chances of great success. In a research situation a leader cannot dictate the ways to do things and how to think. Scientists rely on creativity. A leader helps the team reach its goal. Sometimes this involves being forceful, but more often it involves helping and enabling the team members to do the best that they can.

A team leader should be most dynamic and be visibly initiative in the early stages of bringing the team together and in motivating them then. In this stage, the team members do not understand the goal or their roles. They will be waiting for those to be defined. In the later stages, the team leader must be more circumspect, even when there are tasks to do. Many of the actions, particularly those dealing with negatives, are better dealt with the individuals. This takes more time and effort, but the team dynamics do not suffer as much as they would if issues are deal with in a meeting of everyone. Even praise must be muted or dealt with more subtly if it can be an issue. Some people are easily made envious, for example.

The first task in team building is to create that common goal. This will in all likelihood take some creative people management on your part. The goal of a successfully completed project may not be enough to strongly motivate everyone. The goal, however, can be defined differently for different people. Many people are driven by financial rewards. They work hard in order to receive larger salary increases or bonuses. This, however, may not be true for the majority of scientists. For many researchers the sense of discovery and the curiosity are often enough motivation. Peer recognition reflecting science done well can be motivators for such people. For them financial awards such as bonuses may be enticing and a motivator, but not as effective a one as the recognition through an award that focuses on the discovery of new science and technology. For them what is perceived as a significant award with a small financial reward is much greater than a significant financial one with only a small award or worse still no award at all.

These people are often easily motivated, not by the project goal, but by the technical and scientific problems that must be solved. Encompassing the problems that need solving into the overall goal is only a matter of keeping each stage of the project flowing. In that way the problems are solved in minimal time and the focus moved to the next one. There is little time spent on scientific tinkering around. What is done must be the essentials for the project. If there are variables to define, then there must be both a clear and specific work plan and goal for that stage of the project. The reminder can be that there could be time for add-ons after the project is completed, but it is now a distraction.

Some people are altruistic. It is best for them if their sense of doing the right things and helping can be directly tied to the project goals. This is easy if the project is aimed at something that obviously benefits people such as producing a new pharmaceutical to help treat a disease or monitoring and remediating an environmental cleanup or determining the genetic defects that cause certain illnesses. It is more difficult in that vast majority of times when such benefits are not obvious. If the linkage to benefits cannot be directly done, then these altruistic traits can become at a minimum part of a mentality of helping the team reach success. This can be done by emphasizing the collegial nature of the team, a group of dedicated people working for its goals.

Some people are competitive. They want to do the best job and be recognized for it. They are best motivated if some tangible reward or recognition is a goal. Competitors, however, must be tempered at times so that their individual goals do not become more important to them than the team's goals. This balancing act can be delicate for a team leader to handle. Most competitive people do understand teamwork because their competitive nature has usually been gratified by participation in some activity that involves teams. Sports are the most common pastime of this sort. You, as the team leader, emphasize the needs for teamwork to a competitive person through parallels or analogies to team competitions. These are a natural way of thinking for a competitive person.

People have different values. These may be due to cultural differences. There is a global difference that can influence how teams work. Certain cultures, for instance some of those of East Asian, African, and Middle Eastern cultures, think of a team as a leader with followers. Making people from this type of background comfortable with having initiative and making autonomous decisions can be difficult. Sometimes a person must be taught that learning is not punitive and that trial and error involves errors that are acceptable if they are learned from and arise from thinking while doing a task. Being pusillanimous is ingrained in workers in many cultures (for those of you thinking of rushing to get your dictionary, this word means timid and reserved. It is an intriguing word for me which I have wanted to use in my writings, so please excuse its use if it seems too obscure.) Another problem may arise once people get into being more independent. They might keep testing you on how free they can be and what the boundaries are.

In some strong leader-oriented cultures, followers are expected to diligently do their tasks without questions. In scientific and technical areas, this can be a problem. When there are new tasks or concepts to be learned, teaching and training have to be more interactive. Asking "Do you understand?" will get the culturally driven answer of "Yes" even if it is not true. Not understanding or misunderstanding is not acceptable from a cultural attitude. Lessening this attitude and asking questions for understanding are needed, for example, you might have the person describe the new task or concept in her or his own words. This lets you determine the extent of understanding and can let the other person feel comfortable with asking more.

If you are a person from such a culture and you are put in the role of being a project team leader, you must constantly remind yourself that in order for the team to function well you cannot think of it as your team and run it from a position of authority. This diminishes teamwork and makes everyone in the team your follower who stops thinking and only does what is requested and required. This not only diminishes the potential of the team, but puts much more responsibility and pressure for the success on you. Your job is to help them do the work well by enabling them through the setting of priorities, allocating of resources, funding of budgets, and dealing with any clashes. If you are busy telling everyone their day-to-day tasks, then you cannot also do all of those other things very well.

Dealing with cultural issues is another area that must be covered in a book by itself. Suffice it to say that team members and the team leader must be aware of these differences and learn how to deal with them well in order to have a highly effective team. Some possible resource books are listed in the bibliography.

Protocols and Communications

Modern electronic communications and delivery services make working together at a distance possible, but there are problems that can arise. Relying on e-mail or phone messaging, for examples, requires an acknowledgement of receipt. People can assume that a message sent is one received, but problems do occur. Systems do not always function perfectly, people forget to leave an e-mail autoreply if they are on vacation or a business trip, and other things happen that make the receipt not happen. Requiring a return message acknowledging receipt ensures that things do not "fall between the cracks" in this manner. If actions are required, letting the other team members know that you accept the task and will get it done by a certain date and time are also good practices in far-flung teams.

Renewals and Conflicts

The early stages of a project often are the least troublesome because everyone is enthused and working together still is novel. Once the actual work commences, problems can start occurring, both technically and interpersonally. This is normal. There are only very rare cases of projects working smoothly in both areas. You do not need to be actively involved in either if the project moves forward; if these technical obstacles are overcome and people learn how to work with each other. Be aware of the difficulties, but do not feel compeled to manage every one of them. "Micromanagement" is the derogatory term used for this, being too worried and insecure to not be involved in everything. A team bogs down and the people get less enthused when they feel micromanaged.

If there are more than a few technical difficulties or failures, people may get discouraged. There may be a need to renew the enthusiasm. You can do this by redefining the individual's role in terms of surmounting and solving the new problems through different approaches. Remind people that research involves some failures even when good ideas and experiments are done. If there is an opportunity or the various talents of the people are diverse enough, put together subteams for these specific problems or have them "brainstorm" to give ideas to the researchers working on them. This brings a feeling of more involvement and more doing things that might solve the problem. If people are compatible and have complementary talents then shuffling tasks also is a variation that creates a sense of doing new things.

183

Clashes between people may also happen as they work together. These accumulate until there may be friction and even rancor between people. Part of this may stem from different personality types having to work together. Part of it may simply be that continued close interactions between some people just are not tranquil. The team leader must act as a mediator. This is best done "behind closed doors" with the people involved, individually at first then together. Dealing with negative behavior and personality traits seldom works when done as a group. The individuals feel as if they are being dressed down rather than encouraged to modify to a positive alternative. Once they get that negative perception it will be difficult to get them to listen to your suggestions for change.

Bad People Make a Bad Team – The Chaos of Poor Teamwork

In addition to those bad behaviors that an individual may do that sours the work environment in any circumstances which were discussed in that chapter, certain other bad behaviors are detrimental to good teamwork. They either come out because of the team situation or are magnified by it.

There are several types of people who do not fit well into a team situation because their style of doing work is at odds with their colleagues. A common attribute in them is that they need constant reminding of schedules and deadlines.

Scuttlers are people who try to multitask by doing bits of tasks. They do not plan and do not pick tasks that allow a smooth flow of work. They flit from task to task, not completing any in a cohesive fashion, often loosing track of what needs to be done to keep in synchronization with the others working on the project, and loosing time by being inefficient at their multitasking. This helter-skelter approach seems useful to them, but it is highly inefficient. If the project is a crisis or has tight deadlines, then a scuttler can deter success by not being reliable for his or her portion when it is needed.

Plodders are those people who are not driven. They lack the others motivation and enthusiasm. A crisis may even cause them to shut down waiting for directions. This can be particularly if they defer that critical task for another, but do not tell you that they need direction. They may be waiting on their own spark of ingenuity, but without feeling the pressure of a deadline this can just be avoidance of the tough problem.

Tinkers are scientists who never see a finished task. Every question and every variable that has an effect must be fully studied. Everything has its potential diversions that lead a tinker off on an unnecessary tangent. Each tangent can be a potential project in itself to the tinker's way of thinking. There are never enough repetitions for a tinker to think that the error-bars are small enough.

Scuttlers, plodders, and tinkers all need work plans and schedules. If you give them defined targets, they will work towards them. If not, they work at their own pace. This is true of some other personality types, too. In the team's operation these people can become bottlenecks that slow down successive steps. A team leader must assess which of the team members need these tangible reminders and which do not. Fortunately those that do need them are usually a small proportion of researchers.

Pessimists or *nay-sayers* can be a detriment. They can dampen the motivations of others or create clashes because others tire of their negativity. Some people really are pessimists. Their perspectives are always to see problems and faults. Trying to change their attitudes is difficult. You can, however, reduce or mute the strength of the pessimism.

Some people fall into this pessimist's role because of other motivations. They can be handled by changing the situation to remove those motivations. Some do this when their opinions are not followed. They believe their view is the only possible correct one. They feel slighted and resentful. This may lead them to point out all the failures that occur and presume that these were due to not following their wisdom. These people feel a need to be heard and recognized for their abilities. In the planning stage of any step, ensure that such a person can express his or her ideas. Ensure that these are examined and discussed. Highlight the incorporation of any input, whether the idea itself or the thinking that led to it. This assuages the egotism; often enough to have the person feel involved. If the idea is successful, recognize the person with a "thank you".

On the other hand, unrealistic *optimists* can also be bad. They think things are more easily solved, put forward estimates of time and monies needed that are much lower than is possible, and can wait until there is insufficient time to get things done because they minimize the difficulties. Optimists need to be pinned down on real estimates and reminded that there often needs to be a cushion of time and monies just in case unexpected problems pop up.

Lone wolves are in the habit of doing everything on their own. In a team situation, they can create their own agendas or schedules. They might circumvent you and deal with management or the customer directly. They often will take over their portion of the project and even part of those of others and claim results as a personal accomplishment. A lone wolf does not work well with others unless they are subordinate and directed. Lone wolves are very poisonous on a team as they are virtually unmanageable. If directed by higher management to participate and follow your instructions, they will become minimalists, doing little more than the least amount required, or subversives, undermining your authority, plans, and schedules at any opportunity. It is best to either keep them off the team at the start or to define their role exactly and to as small a reliance as possible.

There is one sort of helping personality that can become an issue in teamwork. If a person is overly helpful to the point of assuming the roles of others or inserting himself or herself into other operations, then there can be a problem if the people being helped feel pushed aside. This usurping of another's role can happen without much effort. If the helpful person assists the other in a task or with a suggestion, the on-going role of the helper being a partner must be mutually agreed to. If not, then the help becomes viewed as meddling and ego clashes occur.

Leadership in a More General Context

Leadership is a rather nebulous concept to most people. A person often is said to have it or not, but pinning people down on what they mean is not easy. The concept is almost as full of ideas as there are people thinking of it. Some of the qualities cited include initiative, dynamism, charisma, and wisdom. My thesaurus (The New Roget's Thesaurus in Dictionary Form, Norman Lewis, Putnam Books, 1978) also lists direction and guidance. In terms of research science, a leader does not guide or direct the technical issues, except maybe in the specific instance of a research advisor in graduate school. A leader guides others to explore and discover new things. This is done by touching on many of the concepts covered in this book, getting people to be curious and ask questions; to get them to review results skeptically; to get them to operate without disciplinary or interpersonal barriers; to get them to be thorough and meticulous; and to get them to aim as one, while thinking as many.

Leaders are said to "Have to make the tough decisions" or "To do what is right". These touch on ethics and on dealing with touchy situations such as dealing with negatives or giving feedback about negative performance. These types of tasks are not easy. They can be deferred or even not done, but the team or organization will suffer in some way. There must be less egocentric and more altruistic attitudes. A leader shares credit with the team, rather than taking it away from the team to become his or hers.

3.9   Balancing Professional and Private Time

The time for one's personal life can be very difficult to fit in or balance with the demands of an active professional career. This is true for business people, attorneys, engineers, architects, doctors, and all of the other professions. There will be time demands for both personal/ private time and professional time. It is easy, especially for the younger scientist at the start of a career, to work ten, twelve, or fifteen hours per day for every day in a week. Younger professors in this situation, focusing on building up a research and teaching program to earn tenure, routinely spend most of their time in their offices or laboratories. The pressures to become established are high. This is very keen for academics because tenure decisions are made after only a few years.

Although it is important to focus on one's career, it should not be the one thing in your life. Being too focused and obsessive can lead to success, but at the high price of there not being anything else. In extreme cases health problems can result. This being said, the consistent message from those in the earlier years of their careers and a good number even of those who have established themselves is how difficult it is to find time to give balance.

Being involved in friends, family, pastimes, sports,

professional and other organizations, charitable causes and organizations, and many other activities add so many benefits. The mental diversions keep you from becoming too involved in the problems of work. This reduces the stress and often even relaxes. Raising a family or doing volunteer charitable work gives perspective – one's daily issues are seen to be less severe and more surmountable. Physical activities are both healthful and entertaining. At a minimum the endorphins must help.

There is some truth to the aphorism "All work and no play make Jack a dull boy". Kathleen Kilway, an organic chemist in the first years of an academic career related that she pushed herself into the habit of working those many hours of long weeks. The tensions and frustrations of building a research program were difficult and so she worked harder. She stopped doing many things that she had been doing for pleasure. She still felt pressured, tense, and frustrated. She then decided to change her schedule to allow regular exercise. She felt less tense and resumed doing other pastimes, such as dancing. They helped her relax and clear her mind some. Her overall happiness and productivity both increased.

Planning and doing certain tasks can help by utilizing time more efficiently and effectively. Subdividing is an important, but underused approach. Look at some of the tasks you must do. Generally people prefer larger blocks of time, a half hour or hour at a minimum. If you rethink and subdivide, that same task may be made up of three ten-minute parts. The use of a schedule and a to-do list can assess the various tasks that need to be done. This finds what amounts of time are needed and how tasks might be divided into smaller segments. This leads to fitting in many small bits of a task into the day and allows for interspersing time for family, personal interests, or friends. This gives a person more balance. Setting up regular schedules of tasks to be done and priority lists defines the most critical work. The time for most tasks can be estimated. A few minutes of planning can make the day's work more efficient and end up with both more work being done and more personal time.

The reading of the literature or correspondence or proofreading reports and manuscripts might be estimated to take several hours per week. Many people think that they must block off this time to both get it all done at one time and to allow for full concentration. This can be cumbersome and difficult to schedule regularly. As with many things, if this reading is not habitual and part of a pattern of behavior, then it is not likely to be sustained over time. It becomes an alternative that "will be done next week when I have more time".

The time it takes to do some tasks, however, can be segmented into many non-continuous bits. For example, reading can be done by each page or section with a bookmark to keep track of progress. With this, five-minute segments while waiting for a meeting to begin, or before any other thing that has a wait before it, can be used to read part of a report or article. A simple folder can be created containing those items that are of the highest priority to be carried and handy for such bits of time. Including a note pad and pen with this folder allows for making notes or memos to oneself.

One professor told me that he did a similar thing by piling new journals on his desk. He was in the habit of glancing through the tables of contents whenever he was put "on hold" during a telephone call. Another planned for the time in airports between connecting flights. He packed a few journals or copies of articles to read while he waited.

The real tricks are in setting up the mental focus to do this in smaller bits of time that fit well within one's schedule and being prepared with material to read in this fashion. Having a pen and paper or some other convenient way to jot down thoughts is also good to prepare for.

A Most Delicate Balance

The time for one's personal life can be very difficult to fit in or balance with the demands of an active professional career. This is the consistent message from researchers. One must make personal time a priority. If not, then professional time dominates. Burn out or personal problems can result.

An organic chemistry professor described how his week was divided into professional time from morning to evening each Monday through Friday. He spent breakfast and dinner with his family. The evenings were also flexibly split between professional and personal interests based on what was actively going on.

Weekends were divided into one day of each. If circumstances made one aspect so that it must take up time normally set aside for the other, then there was a counter-balancing trade-off at another time. For example, if a dinner were missed due to working late, then a few hours in the late morning or early evening extended the personal times.

Another, an environmental chemist, said that while in an industrial position there was pressure to the point of it being an expectation that the work week was 7 days a week of 8 or more hours each. This left no personal balance and so a different position had to be the solution. When that was made, however, he was much happier because his activities and pastimes were important to him both for relaxation and for a sense of self-worth and fulfillment.

There are so many benefits from an active personal and social life that there needs to be little discussion of them. A strong relationship gives both a sense of sharing the joys of success and the commiseration of woes, raising a family gives purpose to the economic benefits of the work and a sense of perspective to any work problems, having interesting hobbies or pastimes give the mind a rest and a sense of fun and accomplishment, doing work for charitable and other groups gives a sense of worth and giving, and being involved in one's religion gives support and reassurance.

4   Career Changes

4   Career Changes

4.1   The Rewards of Working in Industry – Starting and Choosing a Direction from Graduate School to an Industrial Career

Since colleges and universities are where students gain their education, they inherently learn about academic careers through observing and talking with professors. In most cases, however, students do not learn very much about working in industry. There are major differences between the two in the kinds of work done and how it is done. I have written of these in other chapters. There are also major differences in the benefits of working in industry rather than in academia. Since graduate students do not inherently know what these are from their education, I will describe some of them.

Industrial positions generally pay much better than academic ones. Annual salaries are quite higher on the average and pensions and other retirement benefits are as well. There are also bonuses and stock options that on occasion will supplement your salary. Conversely, academia offers job security through tenure while industry does not. A merger, a fall in the company's stock price, and other outside factors may result in cutbacks in staffing or reorganizations in any company, thus leaving any individual vulnerable no matter how skilled she or he may be. If one's organization is eliminated or one's company declares bankruptcy, everyone is at risk, even the most eminent and productive scientists. The up side then is that most companies do give some sort of severance payment and many offer job searching services.

If an individual likes to travel and see the world, an industrial position may be ideal. As an employee of a global petroleum company, one of my former colleagues has traveled to the United Kingdom, France, Germany, India, China, Korea, Thailand, and Malaysia in the past two years, with multiple visits to some of them. His future travels may include Brazil, Russia, and various countries in Africa and South America. Although this high rate of travel is not typical, many industrial chemists do travel since the needs of their companies are scattered geographically with operating plants or customers.

Foreign assignments for months or years are also possible. Thus, being exposed to new cultures and learning other languages are possible for those so inclined. Even if a person does not travel and works only at a company's central or main laboratory research center, at a minimum she or he meets colleagues from all parts of the company. This generally means people from many countries.

One inherent aspect of working in industry is its product focus. These rewards include tangible results that can be directly beneficial or of use to people. A chemist for a pharmaceutical company can work as a team member in the development of a new drug and know from then on that the work saved lives or cured an illness. The personal satisfaction and sense of accomplishment are very strong because of this. Conversely, in some situations a company may be doing business that appears to be detrimental to society through such aspects as pollution and environmental impacts. People working in certain industries find themselves in a defensive posture about their work and employer. Working in the tobacco, petroleum, petrochemical, chemical industries are some that carry this added burden. Circumstances make others notorious.

The technical resources in industry are often much better. Expensive equipment is not rare. An industrial research center may have mass spectrometers, electron microscopes, high-field NMR spectrometers, and other expensive apparatus. Exotic ones, such as a Raman spectrometer, an Auger spectrometer, large-scale distillation or chromatography separators, are found if there is a need. Spending money on reagents, catalysts, and other expensive chemicals is only an issue of need versus benefit.

Most companies offer their products freely or at great discounts to their employees. Although the equivalent monetary amounts would not be much, the psychological ones can be good. For example, a company that makes photographic film and chemicals may have a company store that sells its products at a substantial discount to employees and their families.

An industrial position lets one show that she or he works well under pressure or in adapting to changes. Crises pop up with no warning and the deadlines for solving them are immediate. Projects may be as short as hours in length. A person who can quickly assimilate the background information, assess the situation, create contingencies and alternatives, plan and prioritize tasks to be done, and coordinate efforts will find those talents useful in industry. A dynamic person, one who thrives on adrenaline, is ideal for this problem solver's role.

The opportunities for career changes are much greater in industry. These include assignments in operations, the plants where raw materials and products are manufactured, in marketing and product support where products are sold or where the customers' problems are solved, and in other portions of the company. These other assignments may involve chemical engineering, organic chemistry, toxicology, environmental science, materials science, and many other professional areas.

Another type of career change is away from a technical role and into a business or organizational one. There are few managerial positions in academia, usually only one chairman per department and deans are selected to oversee several departments. There are even fewer opportunities to go into planning, personnel management, or other areas. In industry, there are approximately as many people in these other areas as in the technical ones. A chemist can move into business and finance, personnel areas, and other fields. Even for someone who likes teaching, industry has opportunities as most companies have instructors and trainers in business practices, management skills, communication, and other topics.

4.2   Industry Versus Academia – The Merits of an Industrial Career in Contrast to One in Academia

In a guest editorial in the journal Analytical and Bioanalytical Chemistry (July 2002, 373:209–210) I wrote of the differing needs of industry and academia that a graduate school education should provide. This editorial was significant for me in that it led to the series of essays in that journal that then led to the writing of this book.

Thinking on that topic brings up the natural comparison for a graduate student looking at career options or for a more experienced chemist in one of the two looking at the idea of switching. I will use the basic framework of that editorial to touch on the contrasts between the two types of research work and how each touches on personality differences that make one or the other better. I write this from a presumption that most graduate students have a very good idea of academia from their degree-earning years and from any analogous postdoctoral experiences. On the other hand, their familiarity with the industrial situation is probably very small. I will only touch on the third venue, government laboratories, when it is similar to either or when it contrasts with both.

In aspects other than the scientific and technical advantages, industrial positions generally pay much better than academic ones. Annual salaries are quite higher on the average and pensions and other retirement benefits are as well. Conversely, in industry supplemental income through consulting or giving lectures are rarely possible. In academia, an established, well-known professor may make as much or more as a consultant as the university salary. In addition, patent and other proprietary rights of discovery are retained by an academic (although often shared with the university), while an industrial chemist seldom shares directly in the benefits of holding patents. Gaining successful patents are mainly rewarded through the regular salary administration process as an indicator of good performance or through a bonus system for accomplishments of merit. In both cases the individual does not receive the lucrative rewards of holding a patent and the royalty rights that go with that. This has led to a very large number of professors who have founded their own companies to partake in the business aspects of innovative research and technology.

Now I will move on to the scientific benefits. Industrial analytical chemists work on projects that are defined by the business needs of the company. Projects vary by size, by duration, by technical content, and by application. If a person is good at working on several things over the same time period, setting priorities and scheduling, and dealing with unanticipated crises, then industrial work is a good fit. If a person likes variety in the scientific areas needed to do these projects and enjoys the sense of accomplishment, then industry is a good fit.

The industrial scientist is often the only technical person with knowledge in a particular area of expertise. In industry, the project team usually includes many people trained in other scientific and engineering areas. Each project has its own mix of people, with each member focused on an area of expertise and adding to the team's progress. Diversity of knowledge becomes an inherent part of all analytical work since the others on the team know little or nothing of what analytical approaches are suitable.

As a specific example that I am most familiar with, the

analytical chemist is responsible for both the specific analytical issues surrounding a project, as well as how the proposed solution will mesh with other, often non-technical, factors. For example, a chemical plant cannot often use complex, sophisticated analytical methods because the personnel who would be required to run the method at the plant cannot maintain the equipment adequately nor have the experience to run the equipment effectively. On the other hand, problem solving often utilizes very complex analytical solutions. The range of knowledge of the industrial analytical chemist must be much wider than the academic counterpart, but the requirement for depth of knowledge is not as great. A successful academician is expected to be an expert, even if only in a narrowly defined topic. The industrial chemist must have some working knowledge in a multitude of areas.

The instrumental resources in the analytical laboratories for most large companies usually are far better and more up-to-date than those in a large university. In industry, spending on capital is often based on an annual budget. Each individual chemist justifies their particular needed instruments individually on the needs of the company for that technology. Vast sums are spent each year on new instruments. This is natural since the problems being solved also involve multitudes of money either in cost savings or new revenues. This goes for most resources, as the defining element is the increased money earned by the company relative to the money spent to research and develop an improvement or a new product.

One very well-respected industrial synthetic organic chemist explained that she could not do the enjoyable research she did if she was in academia. "I would spend all of my time writing grant proposals so that my graduate students could have the fun." She has access to separations equipment, molecular spectrometry, catalysts, and a glassware shop that are as good as any in academia.

This contrasts with academia where each individual research professor may be responsible for instrumentation and that must be obtained through inclusion in the research grants. For more expensive items such as a high-field NMR or electron microscope, in industry there is little difference in procurement than for less expensive instruments (a gas chromatograph or UV spectrometer, for instance). In academia, these expensive items are often purchased through targeted research funding or grants that are put forward by several research groups or by a whole chemistry department or even by several scientific departments. This joint ownership often leads to limited access by an individual since there are numerous users.

In academia there is a very strong push for speaking at conferences and publishing. In industry this is much, much less so and very strongly depends on the individual company's policies and attitudes. Some companies encourage these while others strongly discourage them. In some companies, a bright chemist may have fewer than a dozen published papers or presented talks during an entire career. Involvement in the scientific community, particularly in publishing research papers, is less in industry in general. It does depend on the individual and on the policies and attitudes of the company. There are many industrial chemists with a hundred or more publications. In industry, however, involvement is common in groups such as the American Society for Testing and Materials (ASTM) or its counterparts in other parts of the world. This is also true of industry-affiliated associations, such as the American Petroleum Institute, the American Chemistry Council (formerly the Chemical Manufacturers Association), the Ceramics Institute, and other organizations in the United States.

In industry, the diversity of work is usually very broad and changes often with time. New products, projects, and problems arise, then are worked on and completed. A chemist may also work on many different ones in the same period of time. Some may be small and require only hours or days of work, while others may continue for months or even years. The technical areas, the types of chemistry, analysis, and instrumentation vary, as do the application to product quality; process and production problems; regulatory, safety, health, and environmental concerns; globalization issues of uniformity; and other aspects. An industrial chemist may in the morning be extracting soil, water, or plant samples because of a product spill, running those extracts in the afternoon by a particular method, writing up the results that evening, and moving on the next morning to a particular manufacturing plant's processing problem in producing an off-color or smelly product which is a totally different one than what was involved in the spill. Each week and month is filled with different work, different projects.

If a person has no interest in managing or dealing with budgets, then industry offers research positions in which a person only focuses on the science. Many senior researchers commonly still get their hands wet at the bench if they choose to do so (if they do not, then a staff of technicians will perform the research, but it is most often the scientist's choice of which mode). Although it is not common, some scientists can remain as very active experimentalist for their entire careers. This is especially true if the scientist is innovative and creates a reputation of coming up with valuable products or solutions to problems or numerous patents. In academia, having graduate students and writing proposals for and managing research grants are required. After tenure, the expectation is that a professor's research group will grow as the years pass. Thus, she or he is slowly removed from a hands-on active role in research.

Collegial attitudes and relationships in industry also vary widely depending on the individual company's culture and attitude. This can be a plus or a minus depending on one's own attitudes and ways of working with others. Some companies foster collaboration and teamwork throughout their organizations and ways of doing things. This is done through rewarding organizational and team efforts at reaching goals set for groups of people rather than for individuals. Other companies, in contrast, foster an individualistic climate of competition with one's coworkers. Salaries, promotions, and bonuses are given with more of an emphasis on an individual's performance and accomplishments than on team-oriented ones. This often leads the individual to be egocentric. This, in a worse case, leads to increasing personal accomplishments, sometimes at the expense of others. This can be offset by the altruism of being cooperative and collaborative without expectation of recognition or reward. This translates into lesser recognition and fewer gains than the egocentric worker.

Industry offers unique opportunities that contrast greatly with academia in that a multitasking, deadline-oriented generalist has much more of these inherent in the work. Its work is more focused, shorter term, and does not allow as much basic investigation as academia.

4.3   Resume and Curriculum Vitae – Getting the message Across

There are many, many guides on writing resumes. A few are listed in the bibliography. Others can be readily found in many bookstores or through an Internet search. In contrast, there are few sources on the writing of a curriculum vitae (CV). This seems to stem from the more general use of resumes in all venues and in almost all fields (including the non-scientific ones). Business and industry use resumes, not vitae.

A CV, conversely, is only used for job searching in academia and occasionally by governmental research laboratories. There can be both distinct differences in the content of a CV compared to a resume and common elements within the two. Additionally, there is an explanation and some ideas on the complimentary written piece, the cover letter that accompanies the others.

The primary thing to remember in writing any of these three is that they are aimed at getting you noticed enough to stand out above the other applicants. Their purpose is to get you to that next stage in seeking a position, the interview. You do not need to write so comprehensively that you will be offered the position by submitting your resume or CV. You are only using those to sell yourself in person during an interview.

The Resume

A resume according to most of the guidelines is only to be one or two pages in length. If you are a graduate student or a person already working, but within five years of receiving your last degree, then you probably should only have a one-page resume. The exceptional young researcher may have a longer one, but the decision to have a longer resume has to be driven by strong content. Those with more experience can have a two- or more-page resume, but the content defines its length. The resume is a condensed and concise overview of your technical expertise. It is not an autobiography. You should not tell your life story. Only include the essential details that a company wants to know. Personal information, such as birthplace and date, high school attended, marital status, and children, should not be included.

Every technical topical area should be written in an abbreviated form. Often there are also sections dealing with the types of research and work a person has done that would not appear in a CV. These differences include a de-emphasis on work in the published literature and an emphasis on problem solving, working under pressure and with deadlines, and working on project teams that have a variety of team members. These are the sorts of work aspects that are familiar to an industrial hiring person.

Highlight your strongest selling points first. Most companies are inundated with resumes for any advertised position. Their first step is to winnow the chaff out and then more closely look over the remaining wheat. This involves scanning resumes to see which ones meet the minimum requirements. This task is often not even done by one of the technical staff. A person in the human resources department or clerical staff may check for certain words and phrases that signify that the person meets those minimum criteria. The technical staff then only looks at the "better" resumes.

Those that do not meet those will be set aside or filed away. They will not be considered any further. If your qualifications are scattered and spread out through a multi-page resume that contains a lot of irrelevant information (from the hiring person's perspective), then it might not make that first winnowing. Accomplishments that have nothing to do with the position you are aiming for should not be listed. Neither should outdated accomplishments. These only dilute the impact of your stronger, more relevant accomplishments.

On the other hand, items that might connect in some

fashion are good if they can catch a reader's attention. For example, on my first resume, I listed being a member of Mensa as a hobby, knowing that most bright people are curious about that organization of "geniuses". In the course of a half-dozen interviews, it was commented on by interviewers in two of them. The positive images that such other information might generate to differentiate you can be worth the space.

Some models for a resume suggest a summary first. Others suggest bullet lists of accomplishments. In both cases, eye catching statements are read first. Your resume may get less than a minute's notice in the first examination. Brevity in style may include a more clipped wording that does not use pronouns, articles, and the many phrases like "responsibilities included" or "work functions were". For those latter phrases, the listings themselves tell what they are. Be concise. To save space, do not repeat information in different places.

As often happens in industry, proprietary and confidentiality concerns (trade secrets and so on) limit the possibilities of publication. Descriptions of work projects or fields of experience are given instead. Industrial chemists also are, in general, less involved in societies and conferences and receive fewer awards and recognition outside of their own companies. Thus, a CV has these areas while a resume does not. The one area of this sort which does have greater emphasis in industry and is, therefore, a key listing in resumes is the granting of patents. Until recently, patent holding was not highly valued in academia or in government laboratories and so there has been little emphasis on the holding of patents. That attitude has changed in the past decade and CVs commonly list patents now.

When one writes a resume, the target reader must be remembered. It is easy to lose this perspective and write from the information givers view. This, however, results in less effective presentation of oneself. An assessment of a company's need or those needs in a particular technical area should result in a list of items. For example, a position with a petrochemical company that manufactures polymers would have different responsibilities than one in a biotechnology company that manufactures pharmaceuticals. Descriptions of one's experience and expertise should therefore differ in the resumes sent to each type of potential employer.

There have been many stories in the news over the past decade about a lack of full truthfulness in resumes. If this comes to light in the interviewing process, it almost always completely ends any chances. Falsehoods in resumes are so sensitive that many employers, whether in academia, industry, or government, will terminate an employee when these finally come to light. Fabricating accomplishments are neither worth that risk nor the additional on-going risk of damaging one's reputation.

Since resumes are so short, the information in one usually does not need to be changed very often. The two most important items that might necessitate a revision are if a person's contact information changes or if some very significant new accomplishment is done that will draw a positive response from a potential employer. Most on-going accomplishments do not qualify. In normal practice, a resume should be reviewed only either semi-annually or annually to see if it has become dated.

The Curriculum Vitae

A curriculum vitae, in contrast to a resume, is to be a detailed document covering all of one's accomplishments. In addition to some areas included in a resume, such as patents granted, education, the work history, and honors and awards, a CV may also contain the complete listing of a person's publications, oral presentations, and memberships in societies. These latter are often too numerous and mundane for the nature of a resume. Since a CV is so comprehensive and detailed, keeping it constantly current is an option. Whenever any new accomplishment is done such as the joining of a society or the publishing of a paper or presentation of a talk, the CV can be updated. If this is not done, then this information must be noted and kept so that all such changes are added whenever the CV is updated.

My first attempt at putting together a CV was only a few years ago. This was because my career had been spent solely in industry. I received one from a colleague who worked in a government laboratory that I visited to give a lecture. I realized that I should put one together for myself just in case I ever needed one. This one I used as a template and thought it was fine. This first draft was what appeared to be a complete listing of my accomplishments and publications. Briefly, this first CV was a simple, straightforward listing of education, memberships, publications, and honors.

I have since then seen other CVs and have been reminded that a CV is not written for the writer, but rather is intended to be a statement or sometimes even an advertisement for the writer and is aimed at whoever reads it. Potential readers include a person or committee reviewing an application for a position (most commonly in academia a curriculum vitae is used instead of a resume) or a grant reviewer. The CV must be written for those targets. This is one of the common elements of both a CV and a resume; the reader is who it must be written for. The writer must avoid thinking in terms of "This is me, this is who I am" and think in terms of the reader(s) who are thinking "Is this person capable of doing the work that they are applying for?"

Important questions to be asked are "What information in the CV makes the most impact? What is the most unique or noteworthy accomplishment?" These should be presented towards the front with items of descending importance following in a logical order. The reader's energy drops as he or she reads through all of the items in the CV, often resulting in skimming through much of the latter parts or even not reading them at all. The attention span of most readers fades after only a minute of reading, so the writer must increase the readers' interest into reading further into the full details of the CV.

Involvement in societies, especially the holding of offices is a very important attention grabber. So, too, are awards received; invited lectures at conferences, universities, and other research centers; and involvement in special committees. These should be listed early on in the CV in order to garner the most response in the readers' minds.

Thus, a long chronological listing of one's publications may suffice, but splitting it into categories may work better to get the needed message to the reader. For example, classifications such as technical papers, review articles, book chapters and books, patents, and non-technical articles (editorials, book reviews, and letters to the editor are examples) may be used. For a grant proposal, listing patents first may create a very different and much stronger impression as a separate initial section. If some of the patents might be in the same technical area as the research proposed, then this is better than having these items buried among all of the other publications. In this fashion, a review article on the proposed topical area may carry more weight if listed separately. This is especially true if it appeared in one of the more prestigious journals in the field.

Even the splitting of technical papers into topics may be useful if the aim is to impress a grant reviewer. A scientist may have numerous papers, but often only certain ones apply to individual research areas. For example, I have synthesis papers that are of little application if I am proposing research in fluorescence spectroscopy. So listing my fluorescence and organic synthesis publications as separate sections of my CV may keep the reviewer focused on my expertise in fluorescence.

Another aspect of listing of publications (or presentations, positions held, or any other chronological listing) that has differing effects on reading is whether the listing is forward or reverse. That is, do the items proceed from earliest to latest or latest to earliest? Although a forward listing seems to be the more natural, it is probably less effective. The most recent publications have the greater importance for the reader since they are what are more current. This later work is information of the moment, highlighting what the person has done recently or is doing now. The earlier work highlights depth and breadth of knowledge and so also has importance. It, however, does not have as strong an impact on a reader since the reader is looking for the information "What can this person do now related to meet my need or purpose?"

CVs are used both in academia and in government labs. The aims for these may differ. In academia, the two main topics are teaching and research experience. In government labs, research experience is paramount with supervisory experience being sometimes important. Thus, the CV sent to a university must highlight teaching prominently and early in the text. This, on the other hand, would be a distracting irrelevance when sent to a government laboratory.

Inclusion of names and specifics seldom hurt. Giving one's research advisor(s) under the education section may strike a chord with the reader. This is obvious if the research advisor has a well-known reputation. Even if this is not the case, connections cannot be predicted. The reader may have met that professor years ago or even gone to school with him or her. Those linkages make the reader of the CV more receptive. I have even gotten responses from listing my graduate school research topics even though the research is now very dated and seems to me to be rather arcane (the isotopic effects seen in the gas chromatographic retention of some permanent gases. These include questions on other aspects of isotopic effects or adsorption on metals).

For the younger scientist with only a limited number of papers or presentations, the CV may be very short. This can be an opportunity to do something a little different and include more information in the CV. If the title of papers or presentations does not describe the research well, a short synopsis of each item can be added after the typical listing of title, authors, and journal details. These can include more specifics such as the techniques used (MALDI-TOF MS, multi-dimensional NMR, etc.) that may not be evident in the titles. This, of course, gets cumbersome once the number of papers reaches twenty or so, but even for the experienced researcher this technique can be selectively used to explain those papers with ambiguous or less-descriptive titles or to highlight ones of particular note (such as ones receiving a large number of citations).

The Cover Letter

The third important piece of writing that is needed in the application for a position is the cover letter. Whether you are submitting a resume or a CV, a good cover letter complements either. A good cover letter does many things that may help in getting the position. Most job applicants, however, under-utilize the cover letter. A cover letter, at a minimum, states that the accompanying resume or CV is in regard to your interest in a certain position. It should define the position that you are applying for and where you learned of that position. This is because there may often be more than one position open for large companies, government agencies, and even universities. The applications may first pass through a human resources office, secretary, or administrator. Sometimes this first contact with your application is an active one where that person is given criteria to filter out the weaker applicants. A good cover letter that touches on the criteria mentioned in the advertisement ensures that the application survives this filtering.

The cover letter can add a human touch. The cover letter shows interest and enthusiasm. Resumes and CVs by design are factual listings of accomplishment and cannot contain very much of an insight into the personality and attitudes. It is the first that a prospective employer reads. If it is interesting, it can drive the reader to wanting to look more closely at a resume or CV. The cover letter can be tailored towards the criteria that are found in the advertisement for a position.

The cover letter can add information that is specific to the position being sought. The resume is concise and somewhat generic. In contrast, a good cover letter may also be of a similar length or longer, but containing totally different and additional information. The cover letter may express an interest in certain features of the company, such as its type of research work, products made by a company, or its location.

4.4 The Grass is Greener – A Comparison Between Workplaces

The three main types of workplaces – academia, industry, and government laboratories – are often thought of as being distinctly different from each other in their advantages and disadvantages as workplaces. This leads to some mistaken images of each that can then lead a scientist to making career decisions that are not based on realities.

These realities, however, are that all three venues do bear many similarities to each other. Some people might say that they are very similar in that those specific things that are thought of as being highly beneficial or highly detrimental in each are also found in each of the other two. For someone looking at which of these areas to enter into at the start of a career or at a career change from one venue to one of the others, these similarities and differences must be validly assessed. This requires a look at each area and how certain aspects of a position really are within it. That is the purpose of this chapter and thus the title, since the grass is green in each and the shades depend on your personal perspectives.

Anyone assessing careers from the standpoint of bureaucracy, budgets, oversight, and other administrative tasks should understand that none of these areas is immune to them. As is the focus of many of the other chapters, the issues of people skills, the psychological and sociological interactions of people, and communications are also highly important in all three. The following sections, however, will deal with some of those specific aspects that do differ in each venue.

Most of these areas are either highly under-emphasized in graduate school or have no emphasis at all. Professors tend to deal with these issues without involving students. They generally only give a few facts or updates on research grants. Students do not get exposed to bureaucracy, regulations, or funding issues to any great extent. Then they decide on a career direction, get a position, and are hit full force by these issues. At a minimum, it would seem that postdoctoral or very senior graduate students in a research group could both learn from and help with the research professor's efforts in these areas.

There are also several aspects which do differ, but are usually unacknowledged in assessing career direction. One such area is prestige and esteem by friends, family, neighbors, and the public at large. Academicians are highly respected, with an honored role in many cultures. Work in government laboratories generally carries some, especially if you work for an agency dealing with societal issues such as environmental pollution, health and diseases, hunger and nutrition, and others that people see as beneficial. Industrial chemists, in general, gain little esteem by working for a company unless that company is known to provide goods and services that are beneficial. Some industries carry a very negative cachet, such as the petroleum, chemical, alcohol, and tobacco industries.

Funding, Budgets, and Other Financial Considerations

In academia, there once was a time when research funding was competitive, but still relatively easy to obtain for most types of research. If a researcher had a good idea and a well thought-out plan for research to do it, then there was generally a source of funding through the variety of government agencies, foundations, and industry-funded associations. Dealing with budgets and funding was thought of as a bothersome, but necessary, evil within the system. Research proposals were written with some speculation in the research and almost always with more estimated for costs than were really needed or planned.

In the past decade, particularly in the United States, in order to receive funding from government agencies academic research must be much more applicable to commerce and have an impact on the lives of people. There is less possibility of pure research, science done mainly for the interest in discovery of new facts. Within a proposal, the budget estimates are assessed by the reviewers more carefully. Such expenses as those for travel to conferences and "overhead" charges from the institution are not as routinely granted as in earlier times. The research professor must track expenses and support for students since money is tighter.

Funding for "basic research" which aims at understanding of a field rather than at application of its discoveries has dwindled in all three venues. So working in any of them must involve real-world applications or economic benefits as criteria. For those in academia, especially, this has required a different type of thinking and knowledge gathering for writing research proposals. In the past, reliance on the scientific needs of exploring new areas was often enough. Now, the potential benefits to society economically and in the betterment of people's lives must be estimated.

The same concerns of application and value are now as true in government labs as it has been for much longer in industry. In industry, the application timeframe has often shrunk from a few years to only one or two years. This leaves much less time for exploratory experimentation and applications development. In government laboratories the timeframes may be slightly longer, but the pressure to produce useful research is much greater than it has ever been. With this increased pressure to produce tangible benefits, the element of risk-taking has lessened. Research is not only much more applied, but the results aimed at are much more probable at the outset. Failures in experimentation and development are less tolerated.

The lessening of fundamental research in industry, however, is almost to the vanishing point. Many companies think of long-term research as any with a payout of more than one year. Even the applied research is more focused with less time spent on determining all of the details involved in a solution such as the effect of variables or simple permutations that may have future benefits. Working on only the immediate problems is the aim. If those other factors are important, so the thinking goes, then they will be studied if a problem arises. Until then they are not important.

Applicability is also a key factor in government laboratories since funding is approved by legislative bodies that are interested in benefits for the constituencies. The competition among government agencies is also keener as budgets in general have grown smaller. A large proportion of research is now applied, while very little is truly basic or long term. If the application is not immediate, it still is aimed at uses that are less than five years away.

In academia, all funding is through grants of one sort or another. Applying for money is a constant task. Grants usually last for only a year or two, although a few last longer. Seldom do they last more than five years. These grants also are relatively small. They each support the work of only a few students. A large research group may need several simultaneously to accomplish a professor's full program. The professor spends much of her or his time in looking for new money, new sources of it, and writing the grant proposals.

In government laboratories, the mission of the organization is defined. If an idea is not fully encompassed then some selling must be done. This often involves setting up a collaboration with another government agency or a university that is involved in those aspects of the proposed work that do not fall in the scope of your own laboratory or its organization. Internecine fighting for funds among agencies can be very politicized and parochial.

In industry, a proportion of work comes in the assigned form. Tasks with specific goals are given. The researcher comes up with the new technologies, methods, and solutions to meet these goals. On occasion, however, the scientist thinks of an opportunity from knowledge of new things found in the literature or conference or from ideas generated by knowledge of the business needs. This work often must be sold to management in order to get funded. The longer the term the work will take, the more difficult this will be. Also, the more organizations that must be involved or that benefit, the more complicated the marketing will be. More managers will need to be convinced and many will not understand the new ideas and the science that they are based on. This is a difference in industry. In academia and generally in government laboratories, the decision makers understand the science to a good degree.

Monies for doing research are tighter in every venue. Justification of costs and scrutiny of results are both much more than they were and benefits of the proposed research are now also weighed into the decisions. Industrial research in many fields is almost nonexistent, being focused only on customer problems and regulatory issues.

The selling of one's research ideas and plans to a funding source is a task that is involved in all three venues. The tone and aims differ slightly, but marketing is a needed talent for every research scientist. If you are not good at selling your ideas, then your future in each venue will be limited. It is a must in academia, and is a needed talent in the other two if you wish to work on your own ideas. In all three, the key is defining what the funding sources might be and preparing proposals that define the work in terms of their needs.

Personal Gains and Intellectual Property Rights

Universities and academicians a couple of decades ago were not very concerned that the research they conducted did not involve patentable work. Today that attitude is often totally the opposite. Both are very concerned with the patentability of advances in biotechnology areas such as new pharmaceuticals, gene therapy, and recombinant DNA modifying of microbes, plants, and animals for beneficial aims; nanotechnology and other material science; and other breakthroughs with potential lucrative markets. Professors now commonly also are jointly involved in their own companies that deal with the commercial aspects of their research.

This has led to a very different climate on campuses. They are no longer the ivory towered halls of finding knowledge for the joy of discovery. The complications financially and legally can be Byzantine. The movement of professors from one university to another can lead to claims of illegal intellectual property transfer or for portions of royalties due to work done at the first institution. This is just as it can be in the movement from a scientist from one company to another.

If work done by a professor is patented, the university can make a proprietary claim on royalties. This can be beneficial, also. If a professor holds a patent which a company appears to have infringed on, the university's legal resources are large enough to bring suit. For an individual suing a large corporation can be a financial drain. Scientists in this situation, of having their own business and vying with competitors, either means focusing more on the financial and legal issues or bringing in or hiring others that will do that.

One advantage for governmental and industrial researchers is that the responsibilities for discovery and ownership of the intellectual properties are very well defined. Any research that is done which relates to the mission of the government laboratory or to the business areas of the company are not owned in any extent by the individual researcher. The government agencies either retain the intellectual property rights or in some cases they are defined as nonproprietary with an economic benefit to anyone who might use the knowledge. In the case of companies, all rights are owned by the company.

Consequently, in these two venues the researchers rarely can make fortunes from their own research. Some companies will pay an inventor (the title given to a person listed as an originator on many patents) a token amount or give a certificate or plaque. Many companies do not even do this, expecting patents to be part of the work product. In the United States, the restrictions on federal government employees can be so strict that they cannot accept royalties for published work or even a complimentary meal.

Management and Bureaucracy

The common image is that industrial chemists must deal with managers much more than those working in the other two areas. Although this may be somewhat true, academicians and those in government labs have to deal with analogous situations. Academia has numerous faculty committees to deal with the variety of issues. For example, in none of the venues is it allowable to pour any materials down the sink as might have been the practice years ago. Even innocuous and water-soluble samples and chemicals must be disposed of as chemical waste.

Departmental chairs and deans are now often more intrusive because of the greater number of laws and regulations that must be obeyed. Colleges and universities were once immune to labor issues, but now graduate students may be members of a union and be covered by labor laws and collective bargaining agreements. Environmental, health, and safety issues are widespread that cover the use of chemicals and equipment, waste disposal, and the handling of and exposure to hazardous chemicals. Laboratory equipment must follow the local safety regulations, including any regulations covering fire, electrical, or building codes. Unmonitored storage of flammable solvents, haphazard and temporary wiring, and other practices are no longer allowed. The professors and departments can be fined and even found to be criminally negligent.

In each venue any situation that touches issues such as discrimination and harassment must be monitored and dealt with to meet the legal requirements. This has been the case in industry and government laboratories for many years and so there are many procedures that are in place to deal with them. There are mandatory training courses in such issues. Academia is less structured in these areas. Although many universities have offices that deal with any complaints, there is little proactive training of professors and students on correct behaviors.

Along with all the regulatory and legal dealings, there is a corresponding amount of paperwork that must be done and then either sent to an appropriate and defined place or archived. Since few professors have either the status or funding to have their own clerical or secretarial staffs, the burden to do this remains with them or must be done by their students.

Government research is also not immune to these issues that were once thought to be strictly industrial ones. Sample and chemical labeling, training for exposure to hazards and in safe laboratory practices, and proper waste disposal are all now common practices. The practices for performance reviews, salary and bonus administration, and promotion are also similar to those in industry. Mandatory training in safety, waste disposal, and interpersonal topics such as diversity and harassment issues are also common in both venues. Academia is now having to institute some of these as well as universities try to prevent litigation or penalties from governments for safety or environmental noncompliance.

Responsibility

Each career path inherently carries responsibilities that define the position. An academician, for example, will generally be a supervisor since having graduate students is expected. Mentoring and interpersonal skills must be developed to do this. In industry and government positions, being a supervisor and mentoring often happen much later in a career. If a person does not do these things well or chooses not to go that direction in his or her career, it can be structured so that they do not happen at all. From this perspective, an academician bears a large amount of responsibility for her or his success and that of the students in the research group. Someone in the other venues bears much less since the main mission is defined by the organization and much of the supporting activities are borne by it as well.

Generating research funding is inherent to academia. Thus, the scientist must be creative and innovative in areas that research granting groups and government agencies think are worthwhile. This is less so in government labs to some degree since the purpose of the laboratory defines some scientific field of study that is important enough to build and fund the laboratory in the first place. In recent years, however, government laboratories have had to justify their funding much more than in the past. Benefits to the citizens, the economy, environmental protection, or the national defense must be proven.

In many cases, I found that research groups had either greatly expanded their work areas or changed them altogether in order to meet these demands. From my own research area of the analysis of polycyclic aromatic hydrocarbons, three different researchers at different government laboratories in different nations had switched to doing analysis of herbal dietary supplements, detecting the chemical and biological agents used in warfare or terrorism, and determining the marine toxins responsible for shellfish poisoning. Although the former work in PAHs was excellent and of some use, these other research areas were higher priorities because of easier funding or a policy shift for the agencies.

Industry, on the other hand, is very market driven. The aims of most research work are defined by managers rather than the scientists doing the research. On occasion a researcher may think of a new research opportunity or a way to apply known methods to other applications. In this case, the researcher must convince management that the time and resources will have an economic gain. Industrial research almost never is done only for knowledge only. Sometimes this means that the researcher gains the project funding from a sponsoring organization that will be the benefactor. This requires much of the same type of project proposal development as a funding grant in academia or a government laboratory, but usually not with the same level of detail or scientific historical context.

Competition, Parochialism, and Other Interpersonal Clashes

Gaining tenure in academia can be extremely competitive and stressful. A new professor must build up a dynamic research program and find funding for it, attract students, publish in acceptable journals, and become involved in one's division, department, and possibly even in university affairs. All of this must be done without antagonizing the tenured members of the faculty in any way. Even being highly successful in some of these efforts may be bad as far as gaining tenure because jealousy and envy can arise. The stress is very high for the individual because the work structure is built around each professor being very autonomous.

In industry, many of the financial benefits such as salary increases, bonuses, and promotion to higher levels are based on competitive systems. Each employee in essence is competing with her or his coworkers. There can often be unfair distribution of responsibilities in the project work, of the blame with failure, and of rewards and accolades with success. Since very few projects are solely done by one person, this can lead to dissatisfaction, rancor, antagonism, and other negative emotions.

Government laboratories often have similar practices to industry, although seniority may also be a formal factor. If that is the case, it may amount to a quasi-tenure similar to academia with civil service procedures in place. These practices may make transferring to another laboratory or agency difficult, as well as sometimes preventing the hiring of the most qualified people to be coworkers. Government laboratories, however, are very prone to salary, hiring, and promotion freezes if legislatures do not pass budgets. In some cases this may be translated into work disruptions as laws prevent government employees from working without funding.

Job Security and Shifting

This was once thought of as only an issue in industrial laboratories. It has become even more acute in this era of mergers and acquisitions, the subsequent downsizing, and the tenuous nature of venture capital for companies in the new technology areas of biotechnology and materials science. The economic conditions have also highlighted that the work done in academia and government labs is funded mainly through the allocation of government spending. This, in turn, depends on tax income to the governments. In those two types of laboratories, as in industry, hiring for a life-long career is becoming less common. Tenure in a university is no longer a guarantee. In the United States, the euphemism in government laboratories is a "reduction in force" and RIFs are common.

Concomitantly, university positions are so autonomous and dependent on research grants and government labs are so focused on one specific research area (the environment, for example) that diversifying one's expertise is more difficult than in industry. Industrial work normally involves working on many areas and aspects of a problem. In small companies it is a necessity because there are few talented people and many things to do. In larger companies the variety of products and their uses brings more diversity. So industrial chemists may have more tenuous situations, but there is also more expertise in hand to manage the changes. Broadening one's technical base in industry is virtually impossible to avoid.

Additionally, as is covered some in the chapters on moving into supervision, managing, or a non-technical career, government laboratories and industry have more flexibility and career options. In times of job insecurities, finding new opportunities are much greater. Landing a new position in those venues may involve moving away from laboratory research, but that is possible.

The flow between the three areas is higher in the first few years of careers. This is especially true for those who go into academia and either do not gain tenure or who decide that teaching is not part of their goals. Ironically the pattern for entry into academia and industry reverses as years pass. Academia has more rigorous demands on young scientists, often requiring postdoctoral experience, strong communications skills, and a good grasp of the research that is desired for the first four or five years time. Industry seldom is as strict in its criteria for beginning scientists. It is assumed that they do not have the knowledge of a company's product areas or the specific technical experience to deal with the problems associated with them. An "on-the-job" learning is expected.

For experienced scientists, however, few academicians are hired by companies because their research is perceived as being not very applied and "real world". On the other hand, there are many accomplished industrial chemists who transition into teaching and an academic career. Many universities foster this process by offering part-time teaching or research positions.

Administrative Duties Versus Getting in the Laboratory

In talking with scientists from the three venues, this category had the most surprising results. Government laboratories and industry both start out with a certain limited level of laboratory duties. Many institutions and companies have structures and policies that assign some types of laboratory work to those chemists who often do not possess an advanced degree (whether they are called assistants, technicians, or whatever title is used). Advanced degreed chemists do many exploratory or less-routine tasks. For many, this structure continues throughout a career, with only a moderate number of people shifting off to full-time administrative roles.

Academia starts out very differently. A beginning professor does not have graduate or postdoctoral students for at least the first year and often the research group is small for several years. This requires the professor to be involved in every type and level of work. As time passes, however, most professors acquire larger and larger research groups. After a few years of tenure and being established, almost all professors do not do any laboratory work directly. They supervise and direct students who do the research. The professors become strictly administrators.

This seems paradoxical, at first, because it runs counter to our images of each venue. Industry is perceived as management heavy and academia is thought of as the heartland of basic research. These may be true, but only with qualifying

220

statements. Industrial structures have many layers of management between the laboratory researchers and the company's chief executive. These layers, however, supervise and manage wide-ranging parts of the company that may number in the tens of thousands of employees.

In academia, each research group has one head, the professor. So for even the larger research groups of twenty or more students, there is one administrator. The hierarchy in academia does not support this in the same fashion that the higher management levels in a company does. Almost all administrative functions and responsibilities are compartmentalized and localized into each research group.

If you want to still be doing laboratory work throughout your career, academia is the least likely to be a good fit. If you substitute your desire for directly doing research into one of watching others that you direct doing it, then academia is a great fit.

Teaching

Since academia inherently involves a certain amount of both formal and informal teaching, the perception is that this is the only venue in which a person can do that type of work. In addition to the mentoring that is discussed in the chapters dealing with that, scientists in industry and government laboratories teach and train others in science and can have possibilities of doing so in non-technical areas if there is an inclination to doing so. People in those two venues can also teach through colleges and universities in their area and give short courses through societies or conferences.

As one advances in these venues and gains more technical responsibility, the need to train and teach others grows. There will be more technicians working for you. Additionally, there are training opportunities if you develop skills in communications, planning, quality, and other non-technical areas. These can be both in the context of a supervising scientist or in conjunction with the formal training offered by companies or government agencies. Most organizations in these two venues have training staffs. Linking up with them and offering your services to teach certain topics is usually fruitful as those staffs may be shorthanded or be limited in the skills and experience they need to offer many topics.

There is not the same satisfaction in doing these as there is in academia when former students become successful scientists. There is, however, some as former technicians or junior scientists move into more challenging positions or return to graduate school, or utilize those skills learned to be successful on their own projects.

4.5   Changing Hats – Supervising and Managing

For many research chemists, there come points in their careers where alternative types of work become possible. Also, for some their attitudes, inclinations, and talents may indicate that research work is not their best area. They may find teaching, writing, creating computer programs, or other areas much more enjoyable. There are several other professions that are possible for someone with a chemistry degree than only research in the laboratory.

People can change with time. Throughout a person's career, there must be periodic assessments of personality and attitudes, the personal situation, and career direction. This may lead to reassessment of the goals to be aimed for. For example, a young scientist just out of graduate school probably would not have the same personal situation as someone ten years removed from graduate school. Or after several years of successful research, a scientist may not feel as strongly the drive to prove her or his technical capabilities and instead feel more of one to prove management, communications, business, or other skills.

Becoming a Supervisor or Manager

Throughout my couple of decades in an industrial position, I observed many scientists and engineers making the career path change from being a researcher at the bench to becoming a supervisor or manager who sat behind a desk. For a while I even made this journey.

What sorts of things are inherently different in the two areas? The simple answer is just about everything. A successful person in a technical role must be well versed in the scientific areas of the job, capable of keeping up with new developments and trends, be able to clearly communicate technical work, be creative and innovative, and other things technical.

In contrast, a supervisor must be good at interacting with people in various roles, be able to communicate in many non-technical areas, understand budgets and finance, understand safety, health, and labor regulations and issues, have only a moderate technical understanding of the work, among other areas. A manager must be even more versed in finance, business planning, regulatory and legal issues, and needs to have only a passing understanding of the technical areas.

I will discuss each of these contrasts in more detail, but suffice it to say that successful supervisors and managers must have very different skills and personalities than successful scientists and engineers. The types of thinking and skills needed as a supervisor or manager are so inherently different from those needed for scientific success that many good scientists cannot make a successful change or someone who is not an outstanding scientist may become an excellent supervisor and manager or the good scientists have to put in substantial effort at retooling themselves into being supervisors and managers.

Most physical scientists pursue careers in that area because of such qualities as the exploration and discovery, the innovation, for matching theory with experimentation, and the fitting of bits of knowledge into an overall picture of nature. None of these are directly part of the business side where supervisors and managers work. The skill areas there are finance and business practices or the social sciences – the psychology needed to make a team succeed or to deal with subordinates, the sociology of team dynamics and of working with people of different backgrounds.

Many of these skills rely on an understanding of sociological or psychological areas. Most physical scientists (and engineers) do not highly value those areas of study and so are not very versed in them. In order to be a successful supervisor, however, requires a change. The importance of these aspects is shown by the fact that companies and research institutions spent vast sums on training supervisors in those skills.

Some of these are almost anathema to a physical scientist. Budgets and finances are dull and dry; "bean-counting" is the American euphemism, an area of boring and tedious work with no real purpose. In a supervisor's world, however, budgetary issues are paramount. Red or black inks are not just accountants' practice, but real and very important issues in operating an organization. If your organization operates within its budget, you have credibility in cost estimates for new equipment, more employees, or new projects. If your organization is chronically running over budget, you have little leeway or credibility, even in small decisions.

In general, physical scientists either do not inherently have the skills of supervision and managing or find less pleasure in using them than in doing bench work in chemistry, physics, or engineering. They did not choose careers in the social sciences or in business. Of the inclinations for those were strong, the person might have chosen accounting, law, or another career area. This is not to say that a chemist cannot be a good supervisor or manager. The skills are just very different and the opportunity to switch over is not always done because a person is assessed as having strong supervisory or managerial skills before they are given the choice. Much of the time technical success leads to the switch to being a supervisor or manager without any prior assessments of supervisory, business, or managerial skills.

To be a successful supervisor, one must be good at interacting with people in various roles, be able to communicate in many non-technical areas, be able to give criticism in a constructive manner, give praise without appearing to show any favoritism, mediate or deal with conflicts, understand budgets and finance, understand safety, health, and labor regulations and issues, have only a moderate technical understanding of the work, among other areas.

The first rule for supervisors and managers to remember is that those in their organizations are people. Ethical behaviors are part of this. Do not lie or mislead your subordinates to gain your goals. Be honest with them or remain silent.

225

Credibility is an asset for the future. Anything seen as manipulating will be noticed and word of it will soon spread to the others in the organization.

A supervisor soon learns that much of the work involves handling people. Understanding ways of thinking and motivations becomes an important task. The differences in people make it impossible to treat every person the same. What might motivate one person will demotivate or even insult another. Personality differences and cultural influences become factors in determining assignment of tasks, assessments, the creating of project teams, and many other decisions. Much of these are covered in the chapters on personality assessments and teams because those areas are highly impacted by them. For a supervisor, it is sufficient to say that individual variations mean individual ways of dealing with everything if you want to bring to bear the best of each person.

One aspect of being a supervisor that differs greatly from being a bench scientist is interactions with others. In both formal and informal discussions there are different roles and obligations. Being a supervisor, whether it's liked it or not, means to be a representative of the organization. This means that certain things can no longer be the topics of discussion as they might have been before. Discussions of individual's work and private lives cannot be done. Rumors cannot be discussed and spread.

There are both legal and ethical limits on what can be discussed by a supervisor. Those who were once the other collegial bench scientists may now be subordinates. What might have earlier been a discussion of two colleagues on rumors of budget or staffing cutbacks, takes on a much different tone and has a different impact, if it is between a supervisor and a subordinate. Speculation from a supervisor soon becomes real grist for the rumor mill.

Another aspect of the need for changing the types and tones of interaction is in personnel areas. Conflicts of interest in salary administration, promotions, assignments on key projects, and other personnel issues must be avoided. This also prevents the perception of favoritism or bias by others.

Legally, anything done in the supervisor's role relating to an individual should be considered as confidential. Ethically, the supervisor knows much more about each individual and that must be kept privy, as well. If this is not done, the supervisor can become both a legal and personal target. The legal threats may include charges of prejudice or unwarranted disciplination or termination. The personal ones may degenerate into real threats of physical harm.

Any type of negative message is better done in a closed setting, i.e., within an office with closed doors. It should not be done in front of other coworkers. It should not contain personal criticisms unless these are well documented. In that case, there should be involvement of human-resource or other trained people and other levels of management, particularly if there are any disciplinary actions. Documentation of each step and the involvement of these others makes the actions more legally defensible, but more importantly it can lead to improvement without the issue becoming personalized.

A supervisor may have the power to do many things, but these carry responsibilities. Some of those may have repercussions that an individual is better off not shouldering alone. Relying on others, especially in those more alien areas of finance, law, and regulations, has to become part of the way you work. Interpreting labor law or environmental regulations can be a very risky business for someone only trained in laboratory science. Others are professionals in those areas. One must try to remember that outside of the workplace, you do rely on professional attorneys when you are dealing with contracts, wills, and other legal documents and with accountants when you deal with your taxes or estate issues. If it is good practice in your personal life and your company or institution has people doing work solely in those other areas, utilize them.

Letting Go – Delegation and Empowerment

One of the biggest difficulties in transitioning from a purely technical role to the partial one of a supervisor and the non-technical one of a manager is delegation. Relying on the technical skills of others to do work that was once your own is a very basic change in mindset. Instead of planning the work, doing it, and assessing the results, a supervisor or manager must let go of some of the control and have others do the work. Research scientists are doers, so this runs counter to their natures.

Instructing others on how to do laboratory work and then supervising their learning and gradual build up of skills can be trying on one's patience. This is especially true when there are schedules to keep. In the long run, however, a supervisor must let go of some of the direct involvement and responsibility for laboratory tasks. Becoming more of a leader, a coordinator, and a mentor is necessary for a new supervisor or manager. This lets both the workforce and her or him grow into the new roles.

Certain non-technical tasks can and must be delegated, too. This not only frees up the supervisor's or the manager's time to do other tasks; it also helps others gain the skills needed when they move into those roles. The loss of control and the uncertainty of how another person will do a task are issues that the supervisor or manager must overcome. The alternative is involvement in everything, micromanagement, and no one in the organization gain's a sense of responsibility, involvement, ownership, or autonomy.

Delegation involves building a mutual trust that the person can be relied upon and that the supervisor or manager will empower the person to make decisions. This means allowing the person to do as she or he deems best. Alternatives can be discussed. Review of the work that has been delegated may be used as a learning tool, pointing out other options and why they were better or worse than the one chosen. This passes on the thinking behind the decision making.

Part of delegation is empowerment. This is the passing on not only of the tasks and responsibilities to do them, but also the authority to make them happen in whatever fashion is chosen and the trust that the choices will be good ones. These can be iterative and interactive at first, where the person is encouraged to talk about ideas and plans. One key aspect of this is to make sure that everyone involved understands that the tasks are not only delegated, but also supported.

Delivering Messages, Even Bad Ones

As a supervisor and even more so as a manager, speaking to those people within the organization is much different. Personal opinions have to be avoided because one speaks as the representative of that organization. Small comments may have large impacts on attitudes and morale. In the critical times of budget review, when every person is sensitive to talk of cutbacks in funding or staffing, an offhand remark may create worries that effect the operations. Being truthful and assessing the real scope of difficult times is needed, so neither a rosy picture nor a gloomy one are useful in the long run. Another key thing to remember is that people remember and so credibility is another reason to be cautious when speaking.

Negative messages are never easy to pass on, either in the intimate case of a performance review of a subordinate or in the general case of the future of an organization. Telling people negative news can be one of the most difficult things a supervisor or manager must do. Telling people of good news, for an individual of praise in a performance review or of a larger than expected raise or a promotion and for an organization of an increase in funding or the success of a new product or of meeting project targets ahead of schedule and under budget, are all easy to do. The words flow.

Negative news must be said with very carefully chosen words so that the message is not lost through distraction, anger, or fear when the listener hears the bad news. On a performance review, the message should be "This period was not up to what is expected of you, BUT let's talk about why and how we now can improve on those". This is forward looking and does not dwell on the negatives. The same tone must be the aim in presenting any organizational bad news.

Delivering the right messages when times are difficult is another responsibility of managers. Rumors of mergers or reorganizations or decreased budgets add to the general fears of cutbacks. The employees fear for the worst. This can affect morale and productivity, and lead to workplace tensions and conflict. Employees will be stressed and dubious of positive messages as only being a cheery message to keep the flow of work high. Their cynicism should be a factor taken into account in what is said and done.

The message must be positive, but realistic. I will give several examples of what not to do, gleaned from my experiences with a former employer. These are all real things that managers said and did which carried the very wrong message.

Do not only make these pronouncements your only interactions with employees. This sets up the image of the manager only being the bearer of bad news. In addition to formal meetings, take the time to drop in on the offices and laboratories. This should be done at any times, not only the bad ones. This prevents the employees from giving the management the image of being secluded and sequestered, aloof and out of touch with the realities of the work force.

Do not make comparisons of performance of suborganizations in a meeting for the whole. Especially do not make it a listing of sub-organizations that are meeting or exceeding their targets and then one of those that are not. This creates divisiveness, resentment, and competition. If you must make positive reviews of performance, couch them in words that imply that these people or groups are exemplary and with no inferences that others are poorly performing. For the negative messages, make them targeted and deliver them to each group separately.

Do not tell employees that you want to discuss their job situation, future with the company, or career options by leaving a phone message or by sending e-mails. Do this in person. Do this directly, in person, and as immediately as possible. One manager left a phone message that he "wanted to discuss your future with the company", but he was going on a business trip and could not, until a week later. The stress on the employee was huge by being left in a dreaded limbo. If there is a delay, even one over a weekend, then the manager should wait to give the full message. Ideally the manager would realize the importance of promptness and change his or her schedule to talk. To do otherwise is unethical.

Do not delay telling a person that their employment will be terminated so that the work they are doing is completed. This may seem prudent from a work perspective, but lowers the credibility and esteem that the other employees have for the manager. Discuss the situation with the employee. If an extension of the employment is possible for the period until project completion, do everything you can to arrange it.

Do not tell employees that "my job is on the line, too." Their sympathies will not arise. The perception in reorganizations, mergers, and reductions in staff, is that managers always either find a new position or leave with very lucrative severance compensation. Neither of these is likely for the average employee. Trying to paint yourself as being in a similar situation with employees will not work. This rejection of the message is especially stronger when there are more levels between the employees and the manager. Do not compound this by making the first announcements of filling the positions in the new organizational structure that top layer of management unless there are significant changes. If that is done, then the mentality is doubly negative. Morale nosedives because all of the cutbacks must be at the lower levels and the management team's credibility is lost throughout the rest of the process of reorganizing.

Do not continually tell employees every month that the budget is shrinking drastically and then about the possible new projects coming in. After the first few overly optimistic assessments of a different and good future, they become skeptical. Presently probable reasons for optimism and why these are thought to be likely may be useful. These must be tempered to not being a continual part of the messages unless they can be given along with announcements of successes, new work and projects that do make a difference in job security.

These were the actual practices of managers in the analytical chemistry support laboratories in the research center of a large multinational corporation. These managers only learned what they had to perform as they thought was fit. Real capabilities and skills of leadership were not emphasized or taught to them.

The Exalted State of Management

Due to the totally different ways of viewing the world, presentations as management must be different than those a person has given as a scientist to a technical audience. Those focus on experimental accomplishments and problems that were solved, giving details of the science and engineering. To a supervisor (and even more so as a manager) those aspects are minimized and the financial impacts are emphasized. Thus, a scientist may speak of the permutations in experiments leading to a new product or formulation. The supervisor touches on these, but the major thrust of that presentation is the potential new income or savings created by the new product or formulation. A manager, in contrast, may extend this even further to discuss increased market share, decreased operating costs, or other more wide-ranging aspects.

One overlooked skill that is very necessary is in being able to listen to someone. Most people hear the other person speaking, but their thoughts are not focused on the message or the content the other person is giving. The focus is, instead, on what will be the reply to the initial few words or sentences spoken. Being patient and gathering the full message from the speaker often leads to an easier conclusion because the discussion is not sidetracked off of the speaker's topic onto the listener's assumed ones. Real dialogue often avoids problems and also gives the people that one supervises a positive impression that the supervisor understands their concerns.

Not listening often leads to frustration on the speaker's part and if the topic is an issue of some conflict, this can lead to an even more volatile situation. Receiving the speaker's intended message may take time and require questions back to clarify specifics. This can often in itself be a solution to an issue or conflict. Sometime someone only needs to vent, needing someone to hear his or her concerns or frustrations. Sometimes the speaker feels something negative, but in describing it to another he or she gets a better understanding of those feelings and reaches some resolution or level of comfort just by defining the situation.

Extending the list of skills in order to be a successful manager means that one must be even more versed in finance, business planning, regulatory and legal issues, and only needs a passing understanding of the technical areas. Managers often understand little in depth on any project or issue, but must understand all of the aspects of each. This often means relying on the technical, legal, and financial judgments of others. Understanding and assessing this advice and then using it to make decisions are one of a manager's key tasks.

Building a business sense is the greatest new skill area that a manager must have that is different than those used in technical or supervisory positions. This includes understanding the business gains and risks associated with the different decisions and plans that must be made. What are the potential profits for new products taking into account the time, resources, and costs involved? What are the probabilities of each of these potential projects to succeed or fail? What ancillary issues might be involved, such as health or regulatory ones, scale-up issues, or issues of intellectual property, patents, and other proprietary ones?

This business oriented skill set can be thought of as a third one. Technical skills determine if an individual can be a productive scientist or engineer. People skills, such as communication, collaboration, networking, and mentoring, determine if the technically proficient person can become highly successful. Finally, the third set of business skills determine if the highly successful scientist can become a successful manager.

A manager must also understand the linkages between projects, if any, as well as the relative priorities and timeframes for each project. Often when work loads on several projects are in conflict for manpower or other resources, the manager must decide which work gets done by who in what sequence. The focus of a manager's work is on the operational issues of the work group and the technical issues are handled by the individual scientists and supervisors.

Since the purpose of this book is to aid scientists in their research careers, I will not go into anymore on how to be successful once the switch has been made into a supervisory or a managerial career. This brief discussion has only been aimed at letting a scientist know what to expect. Understanding some of the new skills required with those changes and the vastly different roles can help a person decide if those areas are good to get into. A person ill-suited for supervision or managing will not be happy nor be highly successful in them.

There are many, many articles and books on that subject, some of which are listed in the bibliography.

So, in summary, movement into the management ladder, whether in a company, a government laboratory, or even in a university setting (such as in becoming a department chair or dean) involves a shift away from technical roles into ones that are focused primarily on business, financial, and personnel issues. If a person finds that the fire is dying down to do good research, but she or he likes supervising, creating budgets, schedules, and work plans, and running an operation, then becoming a supervisor or manager is a good career path.

4.6   Personal Skills and Assessments

Few physical scientists realize that there is a whole field of study in the social sciences that studies behavioral attitudes and personality traits and how they affect career decisions. In the context of career management and development, these areas can be looked at by their scope as the matching of your personality and attitudes with career, job (position), and task.

     This idea is based on the philosophy that if you are doing it something you enjoy, then you will do a better job. You will also be able to sustain yourself in doing it for much longer. Research science is a creative activity, no different than painting a picture or composing music or writing a book. If your mind thinks of the work as enjoyable and entertaining, the ideas flow easier. If you think of it as wearisome drudgery, then the ideas to come up with for new experiments and theories will not come easily. Enjoyment of the work comes from doing things that match your attitudes and ways of thinking. This is why so many people pursue hobbies and pastimes. The doing of them is voluntary and the enjoyment must be there or other pursuits will soon be done instead. In work we have less flexibility in choosing a career path or profession, but we do have some.

Career Matching

Each individual has a personality. It is built through a wide variety of factors and influences. The traits of it increase or decrease the chances of success in any particular profession. Trying to match personality to the demands of a profession is a key. This is because a successful career will last for decades. A poor fit leads to disenchantment, boredom, and unhappiness. A good fit helps sustain the enthusiasm and energy to do good work throughout one's career.

For experimental science, the basic traits (besides the inherent intelligence to understand the complex ideas that are involved) include curiosity, skepticism, attention to details, and others that are touched on in the sections of this book. Doing good science is as much about attitudes as it is about intellect.

Personality assessments, such as Myers-Briggs, ask numerous questions that help key in on the thinking style and attitudes of an individual. There are four categories that are assessed. Of these, none of the assessments is an absolute. This is because people have few absolutes that they always feel. For example, a person may be able to decide on a menu selection in a restaurant very decisively for a main entrée under normal conditions. If the same person is on an interview in the same situation, she or he may feel that deciding should wait until the hosts have chosen or at least to have asked them what they recommend. These assessments can be used both to find the ideal career and to help teams function better through understanding the similarities and differences in thinking among the members.

One of the better-known and more widely used assessments is the Myers-Briggs Personality Type Indicator (MBPTI). It uses four ranges of temperaments giving an indication of the underlying temperament of the individual. The four are Introverted-Extraverted, Intuition-Sensation, Thinking-Feeling, and Judging-Perceiving. No person has a single and set value for each category. The variables describe behavioral or psychological modes, whether innate or developed. These modes may or may not change over time. Each variable may show a value in one situation that might not be the same as exercised in another set of circumstances.

Introverted (I)-Extraverted (E) - The Extravert gains

energy from interaction with people and social situations. Solitary situations drain their energy. Introverts may feel disconnected in crowds and with strangers, preferring solitary activities and quiet places to restore their energy. While the Extravert is capable of solitary activities, it will not be the preferred mode. In the same sense, the Introvert is capable of being a party animal, but it will be a draining experience after a time.

Intuition (N)-Sensation (S) - The sensation-oriented person most likely describes himself as 'practical', 'factual', 'grounded'. The Intuition person will describe himself as "innovative", "a dreamer". The Sensible person values experience, the wisdom of the past and realism, while the Intuitive person values hunches, the future, and imagination.

Thinking (T)-Feeling (F) - The Thinking person prefers to make choices based on impersonal, objective criteria. The Feeling person prefers making choices based on personal and value judgments. Statistically, there is a sex-based trend in this variable; with 60% of women reported they prefer the Feeling mode while 60% of men reported they prefer the Thinking mode. There is evidence that it is easier for the Feeling person to act in a Thinking mode than for the Thinking person to act in a Feeling mode.

Judging (J)-Perceiving (P) - There is some confusion over the meaning of these terms. The psychologist Carl Jung apparently meant for Judging to mean closure, as in a court judgment closing a case, while Perceiving meant desiring more information and remaining open-ended. Behaviorally, the Judging person takes a deadline as that, makes a decision, and goes on. The Perceiving person resists making the decision, wanting more information to base the decision and feeling uneasy when the decision is made.

The MBPTI is assessed from the answers to numerous questions about feeling and responses for certain situations. For example, how the person behaves when entering a new group of strangers or whether the person likes doing tasks alone or among others. The results are numerical values that describe the range for each variable. There can be midpoint values, which are denoted as an X tendency. The others correspond to strong or weak values for each side of the four variables.

Of the sixteen major categories that result, research

scientists and engineers tend to fall heavily into the INTJ one, although to varying degrees and with some individual traits being different. This is particularly true of men in those professions. Those in one category often find it hard to understand the thinking of others who have a different one.

Job Matching

Much study has been done in this area, particularly for the benefit of companies and institutions that must often fill positions. The main aspect of job matching is personality based. Each institution and company has a culture with a corresponding set of attitudes and values. Someone working in a particular company is expected to hold similar values and attitudes to the other workers there. For example, in some companies it is common for researchers to work late hours and on weekends. Although there are no written rules on schedules, the supervisors may think that someone who sticks strictly to a set work schedule may not be as diligent as is needed.

Some companies are collegial. They have a collaborative mentality that emphasizes teamwork and getting along with your coworkers. The recognition, evaluation, and promotion systems all award teamwork, collaboration, and working without organizational boundaries. Other companies are less so and there are fewer personal interactions. In this culture you are more someone who works in the same place rather than a coworker. These companies highlight an individual's performance and the rewards are given on that basis.

I worked for a company that had a very informal cultural style that also emphasized individuality and less team orientation. After a merger, the employees from the other company who transferred to that work site were easy to recognize. Not only were they less individualistic and more team oriented, their culture had been formally polite. For example, they were used to shaking hands upon meeting, whether at the start of formal meetings, with their immediate coworkers when arriving in the morning, or in a casual meeting in the hallway or when stopping by each other's offices. In speaking with them, they found the new culture colder and more distant.

During mergers, acquisitions, and mass transfers, there

are often little recognition that the different groups of people come from different working cultures. They are just thrust together as part of the new organization charts. If it is a smaller group moving into a large new culture there can be major readjustments in thinking and style. This can lead to resentments, unhappiness, and even a person's leaving because the new culture is too alien. One symptom of this is that the employees still refer to themselves as former employees of the premerger/acquisition companies or refer to how things were done at the former research center. In my former company, a major oil company with a research center in northern California, the various workers referred to themselves as employees of a couple of former merged oil companies and some also as employees of a former separate research center that had been located in southern California. The cultures were so different and unrecognized by the vast majority of "native" researchers that there was lingering resentment.

Another cultural aspect at that research center was in the use of the title "Doctor". It was rarely used. The only noticeable exceptions were if there was a visiting scientist or the older European-educated researchers used it when addressing anyone with a doctorate (and, in certain instances, pointedly did not when referring to someone without one). If you tried to have "PhD" or "DSc" appended to your name on your business card, the request would be returned because there was an unwritten policy against that. Managers said this was to avoid creating class divisions within the company. This was a specious reason since business cards are not exchanged internally, but are given to contacts outside of the company. The common belief, however, was that since the bulk of doctorate holders were chemists and the bulk of managers were chemical engineers with bachelors degrees, at some point the managerial attitude had arisen from insecurity.

Another common, unwritten, behavior is in the dress the workers are expected to wear. In many workplaces, researchers are expected to wear more formal clothes. For the men, this would be coats and ties. Bright colors and more modish styles are not worn and would highlight a different attitude than is acceptable. Women also dress plainly in an analogous way. Some companies' unwritten dress codes are strict enough that it becomes obvious to an outside visitor. If you are interviewing, notice if there is a great similarity in how everyone dresses. This will clue you in on their dress code and hint at other unwritten behavioral expectations. If possible, talk to former workers there to understand the culture.

These policy and cultural attitudes are what differentiates employers. In many industries, the basic technical work, what goes on in the laboratories, does not differ very much from company to company. You should assess your personal style and try to match it best if other considerations are equal when you are choosing. If the other considerations, such as salary, benefits, and location, are factors, then you must understand and accept the atmosphere in your choice. As one unhappy scientist put it "I like the work. It's the job that I hate."

Task Matching

The area that is most overlooked in matching personal skills and the work to be done is the one dealing with the smallest issue, the matching of individual tasks with attitudes. Doing this allows the individual to know the strengths and weaknesses that must be applied to everyday tasks. It gives that person guidance both in assessing areas for improvement and, more importantly, on how to accomplish job responsibilities in an optimal way that relies on strengths.

Self-assessment leads to self-awareness. This then gives a person a better handle of improving and of matching the task to be done to an approach that is optimal.

As is the theme in the previous sections, matching of ones abilities and skills must define the expectations in a specific task or job. Setting goals and expecting one to match them solely on comparison to another person is not valid. For example, if a person is very adept at speaking, then trying to be as good as that individual may lead to failure. If one is not as gifted in some of the natural talents used in speaking, then she or he cannot aim to that goal.

The understanding that a glibness and gift of oratory is not present in everyone does not limit a person. It rather sets a goal customized to the individual's personality. Thinking quickly, organizing one's thoughts, and having a grasp of a broad, but precise, vocabulary are some of those talents. A person must set goals that expand her or his own talents, but on a realistic basis.

240

I once had a colleague who was totally unable to multitask as I was able to and perceived that skill to be. He could work on a single task until it was completed or reached a point in each project where a pause was possible. He could then move to another task. His optimum working mode was to do each task from a start to completion. This meant he might even be able to alternate segments of each project and work on a variety of projects, but in an A1, A2, A3, B1, B2, C1, C2, C3, C4, A4, A5 pattern. Expecting him to be able to have two experiments running at once and to be successful at both was setting an unattainable goal (using the notation this would be A1, B1, A2, C1, B2, A3, C2, B3, C3, A4, etc.).

Having him complete both experiments in his own fashion in a set time period was not possible. So his work schedule was not structured to have him jumping back and forth on simultaneous projects. It was set with looser time schedules that still had targets for completion of parts of projects. This is an example of matching skills to a task.

The Flip Side of the Same Coin

There is a corollary attitude to molding tasks with your own skills as the guide to success. This is to also not expect others to be the same as yourself. The yardsticks for performance should not be comparison to "How I would do it". They should instead be based on how best that individual can do a task given her or his individual ways of thinking, attitudes, and skills. This diversity in approach not only is flexible enough so that it allows each person to optimize their performance, but it can lead to discover new and better ways of doing things that everyone can benefit from. Most people, particularly when supervising or working in a team, must remember that everyone does not think or do work in exactly the same fashion as they might.

In the past decade, especially in the United States, there has been an emphasis in accepting diversity among one's coworkers. Although the common definition is based on the legal aspects of gender, ethnicity, religion, and other bases, in science the added uncommon one of thinking differently must be part of a person's value system.

In order to make this point of differences in thinking being real, I will give a technical example that many readers may relate to. It was one of my first clarifying examples of this. In graduate school, I took a class in group theory and symmetry. It soon became evident that the class was about evenly split into two groups. Some students could readily imagine the molecules and their axes, mirror planes, and centers of inversion by only thinking of them – picturing it in their minds. Others could not and some even had difficulties in seeing those symmetry elements when looking at a molecular model that they held in their hands. The latter group could not be expected to solve the class's problems as easily as those in the first group.

This innate symmetry-assessing talent allowed this first group to perform differently than those in the second group. Inherent differences such as this one highlight the need for a variety of sets of expectations and goals for individuals. Matching talents with capabilities is much more effective than setting rigid criteria for everyone. Expecting everyone to do the same tasks in the same manner is not optimizing the performance of individuals or of teams.

Many traits span a range with the possibility that an individual might be anywhere on that scale. When two individuals interact they might be from opposite ends of the scale and clash. Behavioral scientists define these scales in terms such as individual-collective, global thinking-focused thinking, short term-long term, material oriented-feeling oriented, and many more. Some are not defined as opposite pairs but are scalars, such as deference to leadership and uncertainty avoidance in which the degree of definition of a task is the variable. If you have a very strong attitude on one of these scales you still must accept that other people can be different and that there is no right view. There is yours and others and they differ, but this is not bad. It is diverse.

4.7 Degree of Difficulty – Non-advanced Degree Chemists

The theme of many of these chapters is aimed, or in the least may seem to be aimed, at those chemists with master's or doctorate degrees. I hope that in this section I can help change that impression. Many of the points that are written of, however, are also applicable to those who do not have advanced degrees. Many people working in chemical research and development do not and yet find challenging, creative, and rewarding careers. These include technicians, technologists, analysts, research associates, and others who are also part of the profession of chemistry. A person who does not have an advanced degree still has a potentially challenging and rewarding career in research, but there are many things that must be done in order to increase the involvement and that person's importance in the work.

Even in the initial phases of their careers these individuals provide key roles in scientific research. They perform much of the laboratory work that is needed. Very good laboratory skills are needed, such as meticulousness, observation as experiments proceed, and the abilities to describe the work and results in written or verbal forms. As the career develops, an individual may assume more responsibility with less supervision and guidance. People in this situation can choose either to remain in this career path or of gaining the skills necessary to move into ones occupied by those with advanced degrees. It is a choice that the person can make. Inertia, not making the conscious decision to work towards more knowledge, is a choice for staying in the technician's role. This can be rewarding and filled with challenges, but they are of a different sort than becoming an individual investigator.

Organizations often treat advanced degreed researchers and those who do not have one as if they are two extremely different classes of people. Although there may be specific differences between the two types of people because job responsibilities and career paths have differed, many elements are similar. Collaborative behavior, continuing education, writing and speaking skills, being part of work teams, and many other of the topics described in this book are present for both. The only differences are in the details of what each person has to do, not in the attitudes that lie behind them. That is why I title this chapter "Degree of Difficulty". There are an increasing number of resources for nonadvanced-degreed chemists who wish to grow in their careers. Over the past decade or so, professional societies have acknowledged the importance of such people in the field of chemistry. The American Chemical Society, for example, has a division devoted to their career issues. Special review articles on trends and opportunities appear in its weekly magazine Chemical and Engineering News. Linking up with those resources may give you more ideas on how to build your career.

An area in which these individuals must participate in is continuing education. Learning more increases their career opportunities. If one is satisfied with maintaining one's current situation without regard to advancement, new challenges, or even that coworkers are advancing, then nothing much needs to be done. If, however, one wants to expand and grow in experience then he or she should learn more. For everyone at all levels, there are opportunities in doing work in new technical areas. The barriers to this are limitations on opportunities and the resources to learn. Moving upward in a research organization means taking on more responsibility. Doing this means being able to understand the science better.

For many who do not have an advanced degree, the biggest drawback to advancement or more involvement in the

244

work is a lack of fundamental understanding of the work tasks. They perform experiments but are often not aware of the theories and principles that determine the specific tasks. In many situations, for example, technicians are only expected to perform tasks as directed by a supervising chemist who has either an advanced degree or has much more experience. Sometimes even this is not expected and the person follows a prescribed methodology. Without technical understanding, a person cannot become involved in modification or development of procedures or recognize easily when something is wrong.

A person seeking more knowledge must first assess the resources available. The organization she or he is working in may have in-house resources, such as a technical library and live or recorded instructional classes. A discussion with your supervisor will help you locate these resources and even define a plan to utilize them. Many organizations also support the taking of outside courses. This continuing education may involve taking classes at a local university or through correspondence courses. Another alternative are short courses. Finally, in the least a person can read textbooks or review articles in order to learn the science behind the work.

A structured educational plan must be created in order to both make it a steady high-priority task and to do it in a cohesive fashion to build fundamental and then expert knowledge. The emphasis must be on regular study or work on gaining knowledge and skills. Good intentions might be good for a rare few, but for most people an unstructured learning program starts out with great diligence that eventually fades. The task becomes optional and deferred, with an intention of making up the missed tasks when the schedule becomes less demanding. This rarely happens and the learning becomes spotty. This is an even more likely scenario the more the study is the individual's alone. If there is structure, courses taken, and goals and deadlines set, then the person is forced into habit.

The creation of a structured learning plan is made even easier if this can be done as part of the current job. If that is not possible, the person must look at what goal is being aimed and put together an orderly study program to reach it. What sort of new work or position is being aimed at? What are the duties and skills needed for it? What of these is lacking and by how much?

Answering these define the learning areas. Once these are defined, the resources and options for each learning area can be assessed. Then a plan to learn each can be put together. Do some areas involve learning others first? How much study, formally or informally, can be done? Are there fundamental areas that need to be built up first, such as math skills?

Informal sources include reading books from a technical library. Borrowing from the scientific staff also can provide good sources of books. If you describe your needs and goals, you may even acquire mentors and tutors for the process of learning. If this happens, you might be able to set up short tutorial sessions to augment your reading, which will allow the answering of questions and the passing on of more details and applied, practical knowledge.

Another area in which there are few requirements or opportunities when the work is only performing experiments is that of communication. Any writing that might be required is often simple or informal. There are rarely any demands for speaking. Both of these skills, however, are very necessary in many other career areas, so their lack will limit the possibilities of advancement or career changes. Those doing creative research are judged on their communications skills. Building strong communications skills will allow a person to move into more challenging work. Conversely, a person with poor communications skills will not be taken very seriously in asking for opportunities.

A person must discuss with the supervisor if there are opportunities within the job to improve writing or speaking skills. If there are not, alternatives as part of job training and career development must be pursued. If a structured, on-the-job program is unavailable, then a person should look for outside resources and opportunities. This should include as well any opportunities outside of the workplace. Learning how to write and speak well is so important that anyone desiring to move from traditionally nonadvanced degree work to others must create and take advantage of any possible opportunities.

In many situations there is an initial bias against anyone who does not possess an advanced degree. In some organizations there can even be a bias against those who "only" possess a master's degree and not a doctorate. In such organizations, those with master's degrees are initially not treated similarly to those holding a doctorate. This bias in some instances can be so ingrained that even if a person moves from being a technician into another role, there is a stigma still attached because there he or she does not have an advanced degree.

246

A person needs to continually make others aware of the new capabilities since the perception is that of the former situation where the person only performed laboratory work. This may seem unfair and an additional burden, but others will have an image that must be broken. A new one of more talents and skills must replace it so that the new image has technical respect and is of a peer.

In some cases this is so deeply rooted in an organization that the only recourse for a person seeking advancement is transferring into another organization which either does not have such a stratified mentality or where that person does not have a preconceived reputation. This mentality is not exclusively applied to those who do not have advanced degrees. Some organizations are so stratified that people who start work without an advanced degree cannot assume a larger responsibility even if they earn a doctorate during their employment.

There is some degree of stratification and class both in the ways some organizations operate and in the attitudes of the individuals within it. This, however, generally diminishes to being a non-factor with time. If a person who does not have an advanced degree shows talent, then she or he is recognized for that. This is a corollary to the fact that holders of doctorates are not necessarily skilled at research and will lose whatever aura they have if they are not talented. If your organization seems to operate in a stratified fashion with class consciousness based on degree level, you must decide if there are any real opportunities for you after you work to build your skills. A job change may be necessary in order to fully benefit from a change in education.

Of the three major work areas, industry and government labs present many more opportunities for advancement to nonadvanced-degreed chemists. Numerous examples can be found in any large industrial or government research laboratory of people without advanced degrees that have risen to the same positions as those with advanced degrees.

In these venues there is an initial staffing level based on degree attained. This, however, is supplanted over time by a reliance on capabilities. If the knowledge gained on-the-job allows someone to contribute more, then they move into project roles that match these capabilities. Whether or not a person has an advanced degree is not a criterion, only job performance in these roles is. A talented person who has acquired expertise is often chosen for important assignments and no one thinks anything of whether or not there is a master's or doctorate diploma hanging on their office wall. In fact in many companies, the workforce has a significant number of engineers, most of whom have bachelor's degrees and no advanced degrees. In these companies' cultures being without a doctorate is not a blackmark as even the chairman of the board or chief executive officer may have only the one degree.

Academia still has a large bias in favor of those with advanced degrees. Even those with a master's degree are not treated the same as those with a doctorate. There are positions such as lectureships and support positions, such as in instrument design and repair or glassblowing, which exist for nonadvanced degree chemists in many chemistry departments. Extremely few professorships, however, are open even to very experienced people without an advanced degree. Of these few, even a much, much smaller proportion is in research professorships or in those involving mainly graduate-level teaching. The bulk is responsible for undergraduate teaching and curricula.

The numerous areas discussed in the chapter on non-technical and non-traditional career options are also very good opportunities for those scientists who do not have advanced degrees but who are looking for challenges. Many can be done without attaining another degree, although they generally require some formal training. A few, such as going into patent law, may give the greater opportunity if a specialized advanced degree is also pursued.

4.8 Pursuit of Non-traditional Careers in Chemistry – Chemists Without Lab Coats

The bulk of people who list their occupation as chemist are somewhat similar to the traditional image, someone who runs experiments in a laboratory. This may be in the research mode where new areas of science are explored, but also much lab work involves the testing of materials for quality, environmental analysis, forensics, and other areas of standard testing. In both cases, the chemist actually is a laboratory worker.

There are, however, several career paths in which a person's knowledge of chemistry is not the basis for research or other laboratory work. It can be valuable to know chemistry in many other fields or the chemistry may be done outside of the laboratory. A person can choose these others as their first career choice or can decide later to move into these areas. In the latter, it may be due to finding out that laboratory work is not enjoyable or that better opportunities arise in those other areas. In either case, each of these non-traditional areas requires certain skills and attitudes that are different from research science. The knowledge of science, however, is a fundamental aid in all of these areas.

If you find yourself in either position, a career change may seem daunting. You should remember that a career will be very long, often for thirty or more years. Even if you work in a laboratory doing research for two decades, you can switch careers, spend time learning the new one, and still work in it for a decade.

If one is feeling that research chemistry is not the best career path, then what must he or she do? Career councilors emphasize that a career is a long-duration decision. With that in mind, they rely on skill, attitude, and personality assessments to define the paths. Assessing one's inclinations is covered in the chapter on personal skill and attitude assessments.

These various other work areas also can augment a laboratory career. They can be done part-time to add diversity or to meet specific needs that may arise in one's career. Writing software, writing articles for magazines and newspapers or books, doing statistical analysis, working as an editor for a publishing company, helping on a patent, or consulting readily can fall into this category. If you find that you enjoy these, then their proportion of your work can increase until they are viable in themselves.

Through experience and temperament, each person has inherent advantages and disadvantages in the variety of occupations that are possible. For example, a person who is very uncomfortable in speaking in front of groups would be at a disadvantage as a lecturer. A person who writes well and understands science may become a technical writer. A person with strong computer skills may specialize in writing programs for research chemists to use.

References to some common assessments are listed in the bibliography. In brief, these look at personal attitudes and approaches in thinking. In certain occupations, certain traits will increase the chances of success and long-term satisfaction in one's career. For example, your personality may be one in which interacting with people is a strong skill and a needed aspect of work. Consulting is possible. If you like organization and meticulous and detailed work, running routine assays will be no problem. Neither will being involved in regulatory compliance or data management. On the other hand, if your personality does not have those characteristics, then those career areas will be boring, unchallenging, too steeped in minutiae, and other aspects that will make them bad career choices.

Computational Chemists and Scientific Information and Data Management

Many scientists now use chemistry-oriented software to model molecules or reaction pathways, draw molecular structures, and other tasks. The creation of these programs requires a blend of good chemistry knowledge and strong computer-programming skills. This was once only the domain of a few academicians who had an inclination to do program development as part of their research. With the advent of the personal computer to every scientist, the demand for chemistry programs has grown. Now companies compete with each other for this applications market.

Many uses of today's laboratory work deals with summarized data in electronic form. Databases of thousands of analytical testing results get put into one spreadsheet. This data often must be manipulated or treated by statistical tests. The bulk of people doing this sort of work are computer scientists or statisticians. They often do not understand the origin of the data nor its application. They only gather the information from one set of people, do the manipulations, and pass on the results to others. A scientist who can program or who understands the statistics can make this a much more effective and error-free process.

Science Writing and Journal and Magazine Production

Our world is becoming increasingly one of science and technology. Every part of the world now is connected and each experiences the impacts of new discoveries. New medicines, new ways to communicate, new materials, and many other things developed in the past few decades are now parts of everyone's lives. This creates opportunities for those who understand science and technology to help translate it into forms that non-scientists can understand.

To the average person much of this development is only understood marginally. This has developed into a need for people with communications skills who can make these changes more understandable. Most large newspapers have writers who specifically write articles doing this. The major press services, such as Associated Press, Reuters, Agencie Press Francais, and United Press International, also have writers doing this. Many magazines, such as Scientific American, Science News, and Discover, are aimed specifically at readers wanting to understand today's scientific developments. The electronic and Internet-based media have their own staffs for science reporting.

The range in technical content in science-writing is wide. This means the aims of the writing can be very different and require different styles of writing and in the thinking to create them. For a general circulation newspaper, the intent is to inform the interested public on new discoveries or on the technical background that is the basis for an important issue. It is simplifying complex science into ideas that everyone can understand and must be done without a lot of jargon or scientific language. A different venue is the science-oriented magazines for the general public, such as Scientific American and Science News. These assume a more knowledgeable readership, explain the scientific and technical aspects in much more detail, and get more into the complexities that arise. Finally, there are the more topical writing for scientific magazines and journals like Science, Nature, and Chemical and Engineering News. These are aimed at scientists and try to inform them on a wider range of topics than their own specialized disciplinary focus.

The staff of a technical-oriented magazine or journal includes many more people than the writers. The publication and editorial staffs must be versed in the science and technology that is the content of their publications. They also must be versed in the techniques used to assemble and organize a publication from the various parts. Advertising placement is an example of a specific issue they must deal with. The effectiveness of an advertisement is measured by whoever places it. This determines the rates the publication may charge. Putting an advertisement on different pages can result in differing numbers of readers and respondents. If there will be an article on a particular topic or a review within a particular field, the advertising can be placed so that readers of those also will see advertisements for products or books relating to that area.

Patent Law, Regulatory Affairs, and Other Legal Areas

The increase in technologies has led to an immense number of new scientific discoveries. The globalization of economies and of the scope of businesses has led to the need to uniform legal protections. Traditionally attorneys did much of this work themselves, with aid from technical experts on an as-needed basis. Today the complexity of technical issues, the number of patents, and their eventual huge economic impact makes this way of dealing with patent law obsolescent. The details and intricacies, particularly when comparing one patent to the proposed one, can be difficult for a non-technical person to understand. Conversely, a technical person may not be able to write a patent application in the legal language and terms that cover the nuances of law adequately.

A patent attorney who already is well versed in the intricacies and nuances of science and engineering can write a patent application that both steers around any possible prior claims in earlier patents and lays open a lucrative new technical area that will be protected by the new patent. Often this involves defining any earlier patents in such terms that they in no way overlap with the new technology. The precise wording that this needs involves both legal writing and a strong understanding of the technologies defined in prior patents and in the new work. Attorneys work in a world of black and white where clear-cut and sharp boundaries can, must, and will be created by them. This is a mentality change for most scientists who operate in the multitude of gray shades of a technical world governed by probability.

Attorneys with a chemistry or biology background may work on applications for approval by government agencies that are involved in environmental and worker safety issues. For example, the US Food and Drug Administration has an extensive application process for new products that either are ingested or come in close contact with people. This process involves numerous analytical and toxicological tests to ensure that there will be no significant harm to anyone from the new product. Attorneys must work closely with scientists to gather and assess this information so that it may be included in the application.

Although there are needs for attorneys, those who continue their education by receiving a doctor of law or jurist's degree, there are also needs for scientifically versed people to help attorneys understand the science involved. These people need to become knowledgeable in the law and might even need to pass examinations for certification, but the formal education and time to earn a law degree is not a necessity.

There are an ever-increasing number of laws and regulations. These affect any laboratory. So there are needs in academia and government laboratories, as well as the most-often presumed industrial laboratories, chemical plants, and refineries, for technically qualified people to administer these and oversee operations to ensure compliance. Most of these are in the areas of environmental chemistry, but health effects, exposures, and other toxicological issues are also important.

Regulatory compliance, including such areas as Good Laboratory Practice and the various versions of ISO 9000, are also a career area in which scientists may move into from research. This work involves supervising laboratory operations that do work that must follow those guidelines. Some people also are involved in the training of personnel and in the auditing of such laboratories to ensure their compliance with the guidelines. Having a good laboratory background allows an auditor to question many aspects of the operation that might not be touched on if the person did not have this experience.

Technical Public Relations and Technical Assessment for Business and Investment

In companies, one work area, public affairs, is mainly filled by non-technical people because the work involves specialized knowledge in media relations. Most public affairs people are trained in and have experience as writers for newspapers and magazines or for the electronic media of radio, television, and the Internet. In a technology-based company, however, there is an opportunity for a scientist to do well in this type of work. The base of a technical understanding must be overlaid by experience in communications. The purpose of public affairs is to present specific messages to targeted audiences. It is analogous to the purposes of technical speaking and writing. If you have an inclination and the talents to do the one, you can spread into the broader perspective of communicating for public affairs. Building strong skills in speaking and writing and learning to understand your message and your audience are the things needed.

As an example, in a hotly contested area such as

environmental science, the public's conceptions of a company can be greatly harmed through lack of understandable knowledge. A few "facts" can set their opinions if there are no good counter-arguments. The "facts" that a particular chemical is found in waste water, smokestack discharge, or soil near a chemical facility sounds ominous if also given with only the "fact" that this chemical causes cancer or birth defects. No mention of the levels found versus the harmful levels needs to be made in order to sway the public's opinion against the company. Clearly explaining those specifics in non-technical terms can change that (while one sticking to limits of detection and Ames test doping levels will get lost upon the audience).

With the ever-shrinking budgets of today's world in all three venues, the need for public relations at universities and government laboratories is more than it has ever been. The public is the voting constituency of the politicians who decide on budgeting. If the universities or government laboratories are seen as beneficial to society, they will garner support. If they do nothing, the perception may be that their work is not necessary in economically difficult times.

Investors, stock brokers, and bankers and others from financial institutions that may lend money or carry an initial offering of a company's stock, and managers interested in acquisitions and mergers are among those business people who must deal with science issues while not being versed in them. They all rely heavily upon technical assessments from those who are able to understand the science and technology involved and mesh this with the business and financial issues. This requires technical understanding to assess the science involved, but also for the competing technologies. This merges into the economic and business assessment part, where the advantages and disadvantages are weighed in the technologies, in the legal issues, in the operations of the companies that are competing, and in the target markets they will be competing for. Some understanding of patent law, funding and finances, and assessing companies operations are needed.

Consulting

In today's business climate of smaller work forces, there is an ever-increasing opportunity for this sort of role. Where once many companies could staff experts in all areas, they now have fewer scattered so that few areas are well covered. Even the biggest of organizations can no longer afford to have an expert in every area of expertise. The technical challenges that these companies and government agencies face are of the same or greater difficulties than they ever were. This leads to an opportunity for someone who can provide expertise in the areas not covered in-house.

If a scientist builds up expertise in an area of widespread use, but where there are few other experts, then consulting is an option. In most cases this means a technical niche in which some companies or government agencies do not all have their own people with this expertise. These potential clients of a consultant are in need.

A consultant sells his or her expertise. The strength of this is proven by the publications, patents, and things done in connection with journals, societies, and other activities in the scientific community. Reputation is the most valuable asset. Visibility is another of great value.

A consultant not only needs a keen mind that can absorb the backgrounds of the various projects, rely on an extensive expertise, and come up with innovative solutions, but she or he must be both good at speaking and writing. Being personable helps establish a rapport with clients. Most solutions have to be written in sufficient detail for implementing. This often requires literature searching or other technical supporting evidence.

Two of the key elements for success in consulting are networking and advertising the expertise that is offered so that those who need it are aware. This is not an easy task. Problems that a customer may have will not necessarily be recognized as falling into an area of expertise. Most of the potential clients may not be in your network or even know anyone who is. If you do have an extensive network and a variety of skills, a certain amount of projects will come your way through them if you keep reminding your network of those skills. The solution defines that and this requires that a customer has an idea for a solution.

Reminding people of your skills and looking for opportunities to offer with other new services is a constant task.

Everyone you know in the scientific community may be a link to new projects. You should keep yourself at the forefront of others' thinking. This means corresponding, talking to them by telephone, sending out letters and e-mail notes, and other direct links. This makes your consulting talents a fresh thought. You must get habitual in frequently giving reminders to your network so that when any opportunity arises, you are who they think of and refer potential clients to. Keep expanding your address books and contact listing. Have business cards made up, pass them out frequently, and collect as many from others as you can. Habitually carry some of your own so that even chance meetings become opportunities.

As far as the actual advertising, remember that you offer talents to people with needs. In dealing with potential clients, focus on what their needs are first, then figure how your talents will meet those needs. Too often a scientific consultant focuses on the capabilities and assumes that clients will be looking for those as the solution to a problem. This assumes two fallacies. First, clients are not so technically savvy as to knowing the solution. If they did, then they probably would not be looking for a consultant. You must assume that they do not know. Focus on the application side of your capabilities and seek those with problems in that area. Second, potential clients do not have the time or inclination to seek the best solution. They will go with whatever works in spite of your expertise and thinking being able to create better ones. You must keep your name in their minds for problem solving in general, not just for specific solutions to a particular problem. This way they contact you whenever a problem arises in your general area, knowing that you might be able to help them quickly, effectively, and efficiently.

One aspect of giving value to your clients that touches on your networking and advertising is to always give extra value. What do I mean by this? You want to build a positive relationship with each person in your network and in each client. This will make your clients a part of your network, referring business to you through their contacts and experiences with you.

You do not need to bill your clients for every little effort. If they know the small question that can be answered quickly by telephone or e-mail is a gratuity that you give for their project work, then they will differentiate you from other consultants. Make this known to those who have relied on you and to key people in your network (they may feel reticent to ask you for your opinions and ideas after you become a consultant and you must let them know you are not obliged to charge for every minute).

This policy builds more sustained business, but cannot be used with all potential clients. Some of them will use this as an opportunity to pick your brain, get their needed answers, and not give you any real work. My rule of thumb is that anything less than ten minutes is free and part of selling, but even then I am not totally open with a potential client as far as giving all that they need to know. After all, a consultant is selling knowledge. This is analogous to free samples of a new product, enough to let you try the product and to get you to buy the product.

A second, more technical, area that brings success in consulting is versatility. Being diverse in the expertise you offer means more clients who need one of your skills and therefore more opportunities to link up with them. This entails relying on the knowledge you had before your consulting began and building on that in new applications areas and learning newer ones, as well. For example, there may only be a limited need for an infrared spectroscopist or natural products synthetic organic chemist as a consultant. If the first person expands into environmental remote sensing or the second person does into scaling-up reactions from the bench to batch industrial size, then they expand knowledge into other arenas.

The wider the range of diversity, the more chances of getting projects there are. It is like a restaurant. A limited menu may still be successful if the quality of the food and ambiance are both high. Widening the menu, though, while still maintaining the quality will bring in more patrons. Consulting is a service business, too.

The personal computer makes many aspects of consulting easier. The Internet makes some newer forms of networking and advertising both easier and cheaper. A webpage replaces the needs for mailing out brochures that describe your services. E-mail letters and their analogous electronic newsletters can reach a large audience regularly. Forwarding by the recipients increases the target audience much more. The electronic newsletter contains both advertising for your work and other items that will interest the recipients. If they read for the information, they might read your advertising. At a minimum it puts you into their thinking with each issue. Some possible topics are news items in the field, such as new publications, books, and conferences which your network can help supply you with; a brief item describing something fundamental; sources of interesting compounds; or useful links on the Internet. Putting together a newsletter does take time and effort, but it reminds the recipients of your areas of expertise and gives them something valuable.

With the linkages possible with e-mail, fax, and telephone, you can literally do a project without leaving your office, even if the client is in a far distant country. At other times there may be frequent travel. The clients' needs define the scope of your involvement.

Travel is especially frequent if your consulting includes being an expert witness who gives scientific assessments and opinions in trials. This involves not only strong technical knowledge and credentials that give credence to your testimony, but good speaking skills and the ability to present validation for your client's point of view while also conceding very little to the opposing lawyers. Listening to the other testimony and the precise wording of questions becomes a chess game. You must be skilled in turning the gray areas inherent to science into the sharply defined black and white of the legal world.

The tangible resources needed to be an in-demand consultant are small, but key. A computer with Internet and e-mail access and a fax machine are two. Many companies, particularly in the legal field, still use faxing as a way to transfer information. Brochures, business cards, and a website provide some advertising.

One of the more difficult things to deal with is having access to a good technical library. This helps maintain the expertise and to keep up with all developments in the fields. It also allows for the diversifying into new technical areas and then offering more assistance. It is, therefore, a key. Making arrangements with a local university or company that has a technical library is one of the things that a starting consultant must do. Buying books and subscriptions to key journals help in this, too. These, however, can only be of selective and limited use since the costs are quite high.

The biggest part of the continuing education for a consultant is in time invested in learning. In the laboratory career learning is important. In technical consulting it is vital and the key to long-term success. Keeping up with the literature must become a regular task that involves both maintaining state of the art knowledge in your core areas and diversifying to increase your offerings.

Self-Employed Laboratory Work

In certain cases, there are opportunities for self-employment as a laboratory scientist. Custom synthesis of specific types of molecules and analysis and testing of a specific type may be successful if the right niches are found. The larger companies that already do laboratory work are mainly structured to do work in those areas where there is a great demand. Staffing, equipment, and laboratory space can be expensive. Larger companies, thus, aim for areas where those costs plus a reasonable profit can be earned. Smaller ventures can make do with less and can then be cheaper to use. The niche market is their venue.

Many scientific areas cannot be economical in this business model. An entrepreneurial scientist, however, has one advantage. The expertise and staffing are part of self-employment and the business starter is also the technical staff. Thus the overhead costs of doing the same work are much lower. The producers of custom fine chemicals cannot both have the specific expertise and the lower costs that a smaller operation can have. Thus, there are small companies of essentially one or two scientists who make compounds in a particular class. Analogously, some professors have similar businesses and utilize graduate students as workers (who augment their stipends and assistantships by sharing in the monies made or are paid for their work). Similarly, certain types of analytical work can be offered that no one else can do technically or affordably. Many commercial chemical suppliers and laboratories started out in this fashion.

There are hurdles to overcome. Startup costs in

equipment and chemicals can be large for an individual's budget. Business, safety, and environmental permitting must be done. Many regulations relating to chemicals and hazardous waste must be complied with. Appropriate laboratory space must be rented or purchased.

Additionally, certain types of liability insurance may be a needed cost in running a laboratory. This is also the case in certain types of consulting where the consultant's suggestions may lead to financial risks or expenditures where success is not assured. For example, a recommendation to put in certain process redesigns to solve a product problem has a potential legal responsibility.

5 Bibliography

5 Bibliography

Chemistry Careers

Borchardt J.K. (2000) Career Management for Scientists and Engineers. American Chemical Society/Oxford University Press, ISBN 0-841-23525-2

Day R.A. (1998) How to Write & Publish a Scientific Paper, 5th edn. Oryx Press, ISBN 1-573-56165-7

Feibelman P.J. (1994) A Ph.D. Is Not Enough: A Guide to Survival in Science. Perseus Publishing, ISBN 0-201-62663-2

Friedland A.J., Folt C.L. (2000) Writing Successful Science Proposals. Yale University Press, ISBN 0-300-08141-3

Medawar P.B. (1981) Advice to a Young Scientist. The Sloan Science Series, Basic Books, ISBN 0-465-00092-4

Ramon S. (1999) Advice for a Young Investigator. MIT Press, ISBN 0-262-18191-6

Rodman D., Bly D., Owens F., Anderson A.-C. (1995) Career Transitions for Chemists. American Chemical Society, ISBN 0-841-23038-2

Sinderman C.J. (2001) Winning the Games Scientists Play. Perseus Publishing, ISBN 0-738-20425-0

The Thinking of Science, Serendipity

Gratzer W. (2003) Eurekas and Euphorias: The Oxford Book of Scientific Anedotes. Oxford University Press, ISBN 0-192-80403-0

Sagan C. (1996) The Demon Haunted World: Science as a Candle in the Dark. Pallantine Books, ISBN 0-345-40946-9

Ethics

Ethical Guideline to Publication of Chemical Research (2003) Environmental Science and Technology, Jan 2003, pp 6C-8C

Carnegie D. (1936) How to Win Friends and Influence People. Simon & Schuster, ISBN 0-671-72365-0

Connor J.O., Seymour J. (1990) Introducing NLP – Neuro-Linguistic Programming. Harper-Collins Publishers, ISBN 1-85538-344-6

Covey A.R. (1969) The 7 Habits of Highly Effective People. Simon & Schuster,

ISBN 0-671-66398-4 (hard cover), ISBN 0-671-70863-5 (paperback)

Easler T., Easler C. (2003) The Power to Connect. Creating Communication that Gets Results. Motivated Publishing Ventures, ISBN 0-9732657-0-1

Fisher R., Ury W. (1991) Getting to Yes, 2nd edn. Penguin Paperback USA, ISBN 0-140-15735-2

Harris J. (2001) The Learning Paradox. Capstone Press, ISBN 1-841-12189-4

Harris J. (2002) Emotional Learning. (self-published) ISBN 0-9689-763-5-2

Johnson S., Blanchard K.H. (1998) Who Moved My Cheese? A Parable on Dealing with Change. Putnam, ISBN 0-399-14446-3

Maxwell J.C. (1993) Developing the Leader Within You. Thomas Nelson Publishers, Nashville, ISBN 0-840-76744-7

Maxwell J.C. (1995) Developing the Leaders Around You. Thomas Nelson Publishers, Nashville, ISBN 0-840-76747-1

Maxwell J.C. (1998) The 21 Irrefutable Laws of Leadership. Thomas Nelson Publishers, Nashville, ISBN 0-880-69250-3

Peale N.V. (1952) The Power of Positive Thinking. Ballantine Books, ISBN 0-449-91147-0

Rohm R.A. (1997) Who Do You Think You Are ...Anyway? Personality Insights Inc., Atlanta, ISBN 0-9641080-3-8

Rohm R.A. (1999) Positive Personality Profiles. Personality Insights Inc., Atlanta, ISBN 0-9641080-0-3

Personality Typing for Career Matching and Teamwork

There are a large number of books on this topic, focused on self-assessment, leading teams, or understanding conflicts and their resolution. Many of these authors have other books on these subjects, so if you really are interested look for those other titles.

Barr L., Barr N. (1989) The Leadership Equation: Leadership, Management, and the Myers-Briggs. Eakins Publications, Austin, Texas, ISBN 0-8901-5684-0

Berens L.V., Nardi D. (1999) The 16 Personality Types, Descriptions for Self-Discovery. Telos Publishing, ISBN 0-9664-6244-0

Briggs-Myers I., McCaulley M.H. (1985) Manual: A Guide to the Development and Use of the Myers-Briggs Type Indicator. Consulting Psychologists Press, ASIN 0-8910-6027-8

Briggs-Myers I., Myers P. (1980) Gifts Differing. Consulting Psychologists Press, ISBN 0-89106-011-1 (paperback), ISBN 0-89106-015-4 (hard cover)

Hirsh S., Kummerow J. (1989) LifeTypes. Warner Books, Inc., ISBN 0-446-38823-8 USA and ISBN 0-446-38824-6 Canada

Jeffries W.C. (1991) True or Type. Hamilton Roads Publishing, ISBN 1-8789-0108-7
Krebs Hirsh S. (1985) Using the Myers-Briggs Type Indicator in Organizations. Consulting Psychological Press, Palo Alto, CA. ASIN 9-9980-7114-3
Kroeger O., Thuesen J.M. (1989) Type Talk: The 16 Personality Types that Determine How We Live, Love, and Work. Bantam Doubleday Dell Publishing, ISBN 0-3852-98285-9
Kroeger O., Thuesen J.M. (2002) Type Talk at Work: Determine Your Success on the Job. Doubleday Dell Publishing, ISBN 0-4405-0928-9
Pearman R.R., Albritton S.C. (1997) I'm Not Crazy, I'm Just Not You: The Real Meaning of the 16 Personality Types. Consulting Psychologists Press, ISBN 0-8910-6096-0
Schemel G.J., Borbely J.A. (1989) Facing Your Type. Typrofile Press, Wernersville, PA, ASIN 0-9433-1603-0

Body Language

Dimiytius J.-E., Mazzarella M. (1999) Reading People: How to Understand People and Predict Their Behavior, Anytime, Anyplace. Ballantine Books, ISBN 0-3454-2587-1
Fast J. (2002) Body Language. M. Evans and Co., ISBN 0-871-31982-9 (there have been numerous editions of this book since its original publication in 1970)
Glass L. (2002) I Know What You're Thinking: Using the Four Codes of Reading People to Improve Your Life. John Wiley and Sons, ISBN 0-4713-8140-3

Personality and Working Styles

Bolton R., Bolton N.G. (1996) People Styles at Work: Making Bad Relationships Good and Good Relationships Better. Amacom Press, ISBN 0-8144-7723-2
Brinkman R., Kirschner R. (1994) Dealing with People You Can't Stand. McGraw-Hill, ISBN 0-0700-7838-6

Listening

Burley-Allen M. (1995) Listening: The Forgotten Skill: A Self-Teaching Guide. John Wiley and Sons, ISBN 0-47101-587-3
Kratz D.M., Robinson Kratz A. (1995) Effective Listening Skills. McGraw-Hill, ISBN 0-7863-0122-8

Teams

Duarte D.L., Snyder N.T. (2001) Mastering Virtual Teams. Jossey-Bass, ISBN 0-7879-5589-2

Hackman J.R. (1998) Groups that Work (and Those that Don't) Creating
Conditions for Effective Teamwork. Jossey-Bass, ISBN 1-5554-2187-3
Hackman J.R. (2002) Leading Teams: Setting the Stage for Great
Performance. Harvard Business School Press, ISBN 1-5785-1333-2
Katzenbach J.R. (2000) Peak Performance: Aligning the Hearts and Minds of
Your Employees. Harvard Business School Press, ISBN 0-8758-4936-9
Smith P., Blanck E. (2002) From Experience: Leading Dispersed Teams.
Journal of Product Innovation Management 19(4), July 2002